BASICS OF FOOD CHEMISTRY

BASICS OF FOOD CHEMISTRY

Roshan Chaudhary

ANMOL PUBLICATIONS PVT. LTD.
NEW DELHI - 110 002 (INDIA)

ANMOL PUBLICATIONS PVT. LTD.
H.O.: 4374/4B, Ansari Road, Daryaganj,
New Delhi-110 002 (India)
Ph.: 23278000, 23261597
B.O.: No. 1015, Ist Main Road, BSK IIIrd Stage
IIIrd Phase, IIIrd Block,
Bangalore - 560 085 (India)
Visit us at: www.anmolpublications.com

Basics of Food Chemistry

First Edition, 2008
ISBN 978-81-261-3727-5

PRINTED IN INDIA

Printed at Mehra Offset Press, Delhi.

Contents

	Preface	*vii*
1.	Introduction to Food Chemistry	1
2.	Food-based Vitamins and Minerals	20
3.	Vitamin B-complex	35
4.	Role of Vitamins	59
5.	Role of Minerals	143
6.	Dietary Antioxidants	260
	Index	285

Preface

Food Chemistry is concerned with the study of the physical chemistry of multiphase dispersed systems containing the main functional food components-proteins, lipids, polysaccharides, water and air. Mass spectrometry has been a pre-eminent analytical technique in food science from an early stage in its development. Its development continues to be driven by the innovative application of new experimental techniques and by the insightful interpretation of the information derived using appropriate theories and models.

The scientific aim is to understand how the structural properties of gels, emulsions and foams are related to the interactions between the constituent macromolecules and particles, both in bulk solution and at fluid interfaces, and how the formation and stability of these systems are affected by changes in variables like temperature, pH and flow conditions. The practical objectives for the food scientist are to be able to control the taste and texture of existing food products, and to formulate new structures of high quality using novel combinations of ingredients and processing methods.

Author

Chapter 1

Introduction to Food Chemistry

The diet of any animal contains hundreds if not thousands of different molecules, but the bulk of the ingested nutrients are in the form of huge macromolecules that cannot be absorbed into blood without first being reduced to much simpler and smaller forms - even table sugar (sucrose) cannot be absorbed without first being enzymatically ripped apart into glucose and fructose. The most important enzymatic reaction in digestion of foodstuffs is hydrolysis - the breaking of a chemical bond by the addition of a water molecule.

PROTEINS

Proteins are polymers of amino acids linked together by peptide bonds. Chain length varies tremendously and many dietary proteins have been modified after translation by addition of carbohydrate (glycoproteins) or lipid (lipoprotein) moieties. These modifications will be almost totally ignored in this text. Very short proteins, typically 3 to 10 amino acids in length, are called peptides.

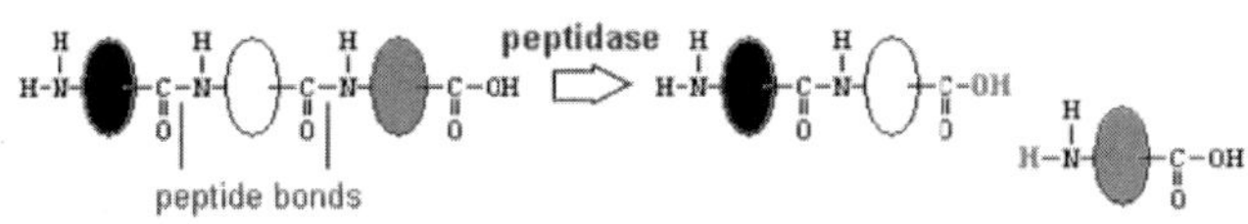

Although very small peptides can be absorbed to a limited degree, for all intents and purposes, proteins must be reduced to single amino acids before they can be absorbed. Enzymes that hydrolyze peptide bonds and reduce proteins or peptides to amino acids are called *proteases or peptidases.*

LIPIDS

Fatty acids are present in only small amounts in animal and plant tissues, but are the building blocks of many important complex lipids. True fatty acids possess a long hydrocarbon chain terminating in a carboxyl group. Nearly all fatty acids have an even number of carbons and have chains between 14 and 22 carbons in length.

The principle differences among the many fatty acids are the length of the chain (usually 16 or 18 carbons) and the positions of unsaturated or double bonds. For example, stearic acid (pictured below) has 18 carbons and is saturated.

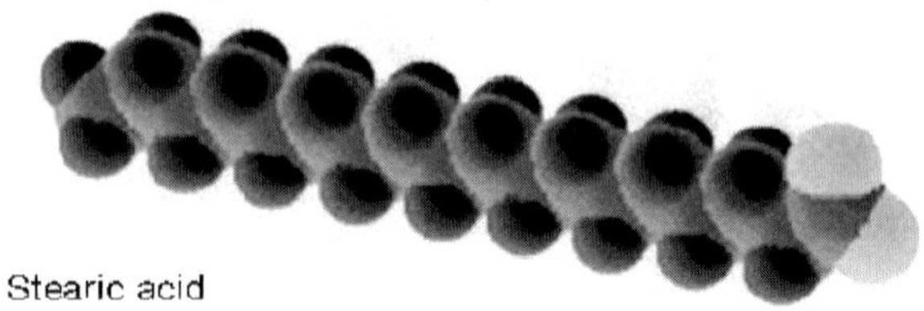

The so-called "short-chain" or volatile fatty acids are 2 to 4-carbon molecules of great importance in intermediary metabolism and as the mainstay of ruminant nutrition. They are represented by acetic, butyric and proprionic acids.

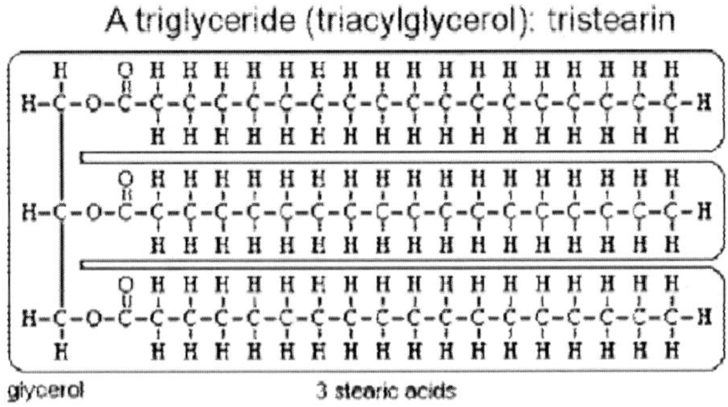

Fig: A triglyceride (triacylglycerol): tristearin

The most abundant storage form of fat in animals and plants, and hence the most important dietary lipid, is neutral fat or triglyceride. A molecule of triglyceride is composed of a molecule of glycerol in which each of the three carbons is linked through an ester bond to a fatty acid. Triglycerides cannot be efficiently absorbed, and are enzymatically digested by pancreatic lipase into a 2-monoglyceride and two free fatty acids, all of which can be absorbed. Other lipases hydrolyse a triglyceride into glycerol and three fatty acids.

CARBOHYDRATES

The diversity of dietary carbohydrates necessitates discussion of several classes of these molecules, ranging from simple sugars to huge, branched polymers.

Monosaccharides or simple sugars are either hexoses (6-carbon) like glucose, galactose and fructose, or pentoses (5-carbon) like ribose. These are the breakdown products of more complex carbohydrates and can be efficiently absorbed across the wall of the digestive tube and transported into blood.

Disaccharides are simply two monosaccharides linked together by a glycosidic bond. The disaccharides most important in nutrition and digestion are:

- Lactose or "milk sugar": glucose + galactose
- Sucrose or "table sugar": glucose + fructose
- Maltose: glucose + glucose

Oligosaccharides, which include disaccharides, are relatively short chains of monosaccharides which typically are intermediates in the breakdown of polysaccharides to monosaccharides.

Polysaccharides are the most abundant dietary carbohydrate for all except very young animals. You should be familiar with three important polysaccharides, each of which is a large polymer of glucose:

- Starch is a major plant storage form of glucose. It occurs in two forms: alpha-amylose, in which the glucoses are linked together in straight chains, and amylopectin, in which the glucose chains are highly branched. Except for the branch points of amylopectin, the glucose monomers in starch are linked via alpha glycosidic bonds, which, in the digestive tract of mammals, are hydrolyzed by amylases.

- Cellulose is the other major plant carbohydrate. It is the major constituent of plant cell walls, and more than half of the organic carbon on earth is found in cellulose. Cellulose is composed on unbranched, linear chains of D-glucose molecules, linked to one another by beta glycosidic bonds, which no vertebrate has the capacity to enzymatically digest. Herbivores subsist largely on cellulose, not because they can digest it themselves, but because their digestive tracts teem with microbes that produce cellulases that hydrolyze cellulose.

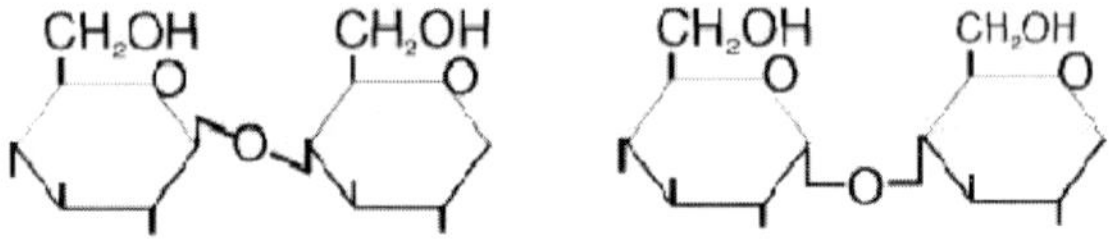

beta (1 → 4) bond
Cellulose

alpha (1 → 4) bond
starch, glycogen

- Glycogen is the third large polymer of glucose and is the major animal storage carbohydrate. Like starch, the glucose molecules in glycogen are linked together by alpha glycosidic bonds.

DIGESTIVE SYSTEM

Consider for a moment a Big Mac. The purpose in your eating a Big Mac, other than simple hedonism, is to assimlate the nutrients it represents and make them available to build, repair and maintain your own tissues, as well as provide energy for studying and occasional other pursuits.

You may have asked yourself - *"Exactly what nutients are present in a Big Mac that I can assimilate?"* MacDonald's comes close to full disclosure in this regard, but what they don't tell you is that in order to take advantage of these nutrients, you have to provide the means to carefully break them down into much smaller molecules that can be imported into blood. Luckily, your digestive system takes care of this very complex process so efficiently that most of the time you don't even need to think about it.

At its simplest, the digestive system it is a tube running from mouth to anus. This tube is like an assembly line, or more properly, a dissembly line. Its chief goal is to break down huge macromolecules (proteins, fats and starch), which cannot be absorbed intact, into smaller molecules (amino acids, fatty acids and glucose) that can be absorbed across the wall of the tube, and into the circulatory system for dissemination around your body.

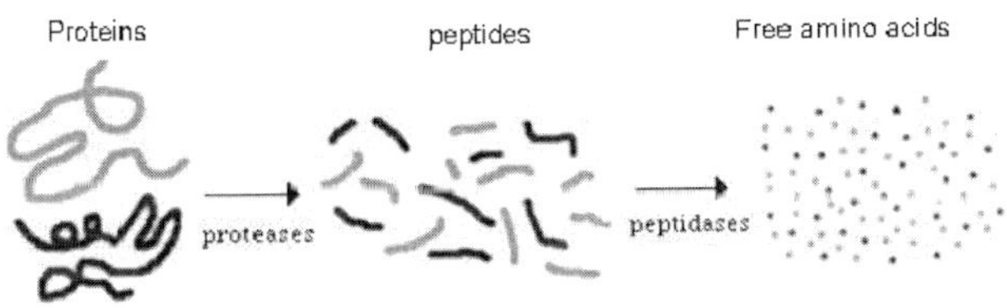

The breakdown of foodstuffs like a Big Mac is accomplished through a combination of mechanical and enzymatic processes. To accomplish this breakdown, the digestive tube requires considerable assistance from accessory digestive organs such as the salivary glands, liver and pancreas, which dump their secretions into the tube. The name "accessory" should not be taken to mean dispensible; indeed, without pancreatic enzymes you would starve to death in short order.

In many ways, the digestive system can be thought of as a well-run factory in which a large number of complex tasks are performed. The three fundamental processes that take place are:

- *Secretion:* Delivery of enzymes, mucus, ions and the like into the lumen, and hormones into blood.

- *Absorption:* Transport of water, ions and nutrients from the lumen, across the epithelium and into blood.
- Motility: Contractions of smooth muscle in the wall of the tube that crush, mix and propel its contents.

Each part of the digestive tube performs at least some of these tasks, and different regions of the tube have unique and important specializations.

Like any well-run factory, proper function of the digestive system requires robust control systems. Control systems must facilitate communication among different sections of the digestive tract (i.e. control on the factory floor), and between the digestive tract and the brain (i.e. between workers and managment).

Control of digestive function is achieved through a combination of electrical and hormonal messages which originate either within the digestive system's own nervous and endocrine systems, as well as from the central nervous sytem and from endocrine organs such as the adrenal gland. Different parts of these systems are constantly talking to one another. The basic messages are along the lines of *"I just received an extraordinary load of food, so I suggest you get prepared"* (stomach to large intestine) or *"For goodness sake, please slow down until I can catch up with what you've already given me"* (small intestine to stomach).

Finally, a note about differences in digestive anatomy and physiology among animals. The digestive systems of humans, dogs, mice, horses, kangaroos and great white sharks are, to a first approximation, virtually identical. If you look more carefully however, it becomes apparent that each of these species has evolved certain digestive specializations that have allowed it to adapt to a particular diet.

These differences become particularly apparent when you compare a carnivore like a cat with a herbivore like a goat or a horse. Goats and horses evolved from ancestors that subsisted on plants and adapted parts of their digestive tracts into massive fermentation vats which enabled them efficiently utilize cellulose, the major carbohydrate of plants.

In contrast, cats evolved from animals that lived on the carcasses of other animals, and have digestive systems that reflect this history - extremely small fermentation vats and essentially no ability to utilize cellulose. Bridging the gap between carnivores and herbivores are omnivores like humans and pigs, whose digestive tracts attest to a historical diet that included both plants and animals. The image above shows a young omnivore in the company of herbivore and carnivore friends.

ANATOMY OF THE DIGESTIVE SYSTEM

The digestive system is composed of the digestive or alimentary tube and accessory digestive organs. The basic terminology used to describe parts of the digestive system is shown below and more detailed description of each is presented in later sections.

The digestive system depicted above - a carnivore - is the simplist among mammals. Other species, even humans, have a more or very much more extensive large intestine, and ruminants like cattle and sheep have a large set of forestomachs through which food passes before it reaches the stomach.

Each of the organs shown above contributes to the digestive process in several unique ways. If you were to describe their most important or predominant function, and summarize shamelessly, the list would look something like this:

- *Mouth:* Foodstuffs are broken down mechanically by chewing and saliva is added as a lubricant. In some species, saliva contains amylase, an enzyme that digests starch.
- *Esophagus:* A simple conduit between the mouth and stomach - clearly important but only marginally interesting compared to other regions of the tube.

- *Stomach:* Where the real action begins - enzymatic digestion of proteins initiated and foodstuffs reduced to liquid form.
- *Liver:* The centre of metabolic activity in the body - its major role in the digestive process is to provide bile salts to the small intestine, which are critical for digestion and absorption of fats.
- *Pancreas:* Important roles as both an endocrine and exocrine organ - provides a potent mixture of digestive enzymes to the small intestine which are critical for digestion of fats, carbohydrates and protein.
- *Small Intestine:* The most exciting place to be in the entire digestive system - this is where the final stages of chemical enzymatic digestion occur and where almost almost all nutrients are absorbed.
- *Large Intestine:* Major differences among species in extent and importance - in all animals water is absorbed, bacterial fermentation takes place and feces are formed. In carnivores, that's about the extent of it, but in herbivores like the horse, the large intestine is huge and of critical importance for utilization of cellulose.

MICROANATOMY OF THE DIGESTIVE TUBE

Remarkably diverse and specialized processes take place in different sections of the digestive tract, but there is a fundamental consistency in the architecture of the tubular digestive tract. With few exceptions, the wall of the digestive tube from the mouth to the anus is composed of four basic layers or tunics.

Tunica serosa is the outermost covering of the digestive tube. In most of the digestive tract (stomach and intestines) it consists of a thin layer of loose connective tissue covered by mesothelium (a type of squamous epithelium that lines body cavities); within the peritoneal cavity, this structure is also referred to as visceral peritoneum.

In the abdominal cavity, the serosa on each side of the tube fuses together to form a suspensory structure called mesentery, which houses vascular and nervous supplies to the digestive tract and is continuous with the lining of the cavity. In regions outside of the abdominal cavity where the the digestive tube is essentially affixed to adjacent structures via its outer layer of connective tissue (esophagus and rectum), this tunic is referred to as *tunica adventitia* instead of tunica serosa.

Tunics

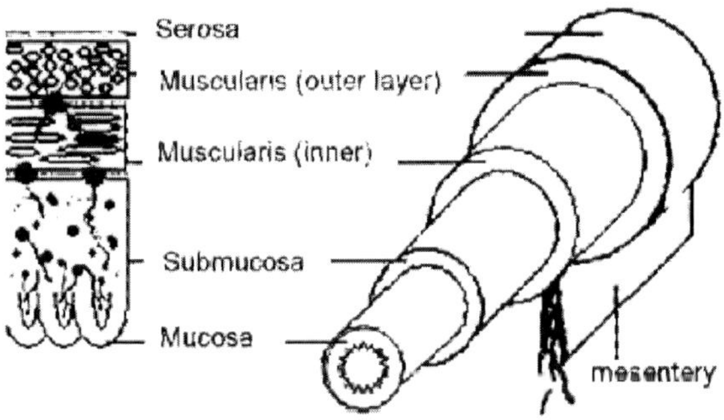

Tunica muscularis endows the digestive tube with an ability to be motile. In most of the digestive tube, this tunic consists of two thick layers of smooth muscle. Muscle fibers in the inner layer are aligned circularly, whereas those in the outer layer have a longitudinal orientation.

This combination of circular and longitudinal smooth muscle gives the tube an ability to perform complex movements that squeeze and propel ingesta in the lumen. Between the inner circular and outer longitudinal layers of smooth muscle is another critical component of the digestive tract's nervous system - the myenteric plexus.

Tunica submucosa lies immediately beneath the mucosa, and is a layer of loose to dense connective tissue containing blood and lymphatic vessels. The submucosa also contains the submucous plexus, a critical component of the digestive tract's nervous system which provides nervous control to the mucosa.

Tunica mucosa is the innermost layer of the digestive tube and lines the lumen. *Among the four tunics, the mucosa is most variable in structure and function, endowing the tube with an ability to perform diverse and specialized digestive tasks along its length*. Of critical importance in this regard are the epithelial cells that cover the mucosa and are thus in direct contact with the lumen. This epithelial cell sheet (*lamina epithelialis*) is distinctly different in different regions of the tract. Indeed, in most of the tract, several different cell types contribute to the epithelium, including cells dedicated to secretion, absorption or production of hormones.

These distinctive differences in architecture of the epithelium can be seen below in the micrographs of mouse digestive tube. The magnification of all four images is identical and the epithelial layer is oriented toward the top.

Beneath the epithelium, but still within the tunica mucosa is a layer - the *lamina propria* - of loose connective tissue through which course blood vessels and lymphatics that supply the epithelium. This layer also contains lymphatic nodules important to immune functions of the digestive tract. Finally, beneath the lamina propropria is a thin layer of smooth muscle (*lamina muscularis mucosae*) which permits the mucosa to dynamically move and fold.

MICROBIAL LIFE IN THE DIGESTIVE TRACT

The gastrointestinal tract contains an immensely complex ecology of microorganisms. A typical person harbors more than 500 distinct species of bacteria, representing dozens of different lifestyles and capabilities. The composition and distribution of this menagerie varies with age, state of health and diet.

The number and type of bacteria in the gastrointestinal tract vary dramatically by region. In healthy individuals the stomach and proximal small intestine contain few microorganisms, largely a result of the bacteriocidal activity of gastric acid; those that are present are aerobes and facultative anaerobes. One interesting testimony to the ability of gastric acid to suppress bacterial populations is seen in patients with achlorhydria, a genetic condition which prevents secretion of gastric acid. Such patients, which are otherwise healthy, may have as many as 10,000 to 100,000,000 microorganisms per ml of stomach contents.

In sharp contrast to the stomach and small intestine, the contents of the colon literally teem with bacteria, predominantly strict anaerobes (bacteria that survive only in environments virtually devoid of oxygen). Between these two extremes is a transitional zone, usually in the ileum, where moderate numbers of both aerobic and anaerobic bacteria are found.

Table: Microbial Populations in the Digestive Tract of Normal Humans

	Stomach	Jejunum	Ileum	Colon
Viable bacteria per gram	$0 - 10^3$	$0 - 10^4$	$10^5 - 10^8$	$10^{10} - 10^{12}$
pH	3.0	6.0-7.0	7.5	6.8-7.3

The gastrointestinal tract is sterile at birth, but colonization typically begins within a few hours of birth, starting in the small intestine and progressing caudally over a period of several days. In most circumstances, a "mature" microbial flora is established by 3 to 4 weeks of age.

It is also clear that microbial populations exert a profound effect on structure and function of the digestive tract. For example:

- The morphology of the intestine of germ-free animals differs considerably from normal animals - villi of the small intestine are remarkably regular, the rate of epithelial cell renew is reduced and, as one would expect, the number and size of Peyer's patches is reduced.
- The cecum of germ-free rats is roughly 10 times the size of that in a conventional rat.
- Bacteria in the intestinal lumen metabolize a variety of sterols and steroids. For example, bacteria convert the bile salt cholic acid to deoxycholic acid. Small

intestinal bacteria also have a important role in sex steroid metabolism.

Finally, bacterial populations in the large intestine digest carbohydrates, proteins and lipids that escape digestion and absorption in small intestine. This fermentation, particularly of cellulose, is of critical importance to herbivores like cattle and horses which make a living by consuming plants. However, it seems that even species like humans and rodents derive significant benefit from the nutrients liberated by intestinal microorganisms.

FOOD BIOTECHNOLOGY

We have used biotechnology to manufacture food products for more than 8,000 years. Bread, alcoholic beverages, vinegar, cheese and yogurt, and many other foods owe their existence to enzymes found in various microorganisms. Today's biotechnology will continue to affect the food industry by providing new products, lowering costs and improving the microbial processes on which food producers have long relied.

Many of these impacts will improve the quality, nutritional value and safety of the crop plants and animal products that are the basis of the food industry. In addition, biotechnology offers many ways to improve the processing of those raw materials into final products: natural flavors and colors; new production aids, such as enzymes and emulsifiers; improved starter cultures; more waste treatment options; "greener" manufacturing processes; more options for assessing food safety during the process; and even biodegradable plastic wrap that kills bacteria.

Improving the Raw Materials

The first generation of transgenic crops primarily benefited farmers. Although there are consumer benefits in growing these crops, the benefits are largely invisible to consumers. For example, studies have shown that because insect-resistant corn (Bt corn) sustains relatively little insect damage, fungi and molds cannot infect those plants as easily as non-insect-resistant crops. Therefore, the level of toxins, such as aflatoxin, produced by these pathogens, some of which are fatal to livestock, is much lower in Bt corn than non-Bt corn.

The benefits of the next wave of biotechnology crops will be more obvious to consumers. Some of those benefits will involve improvements in food quality and safety, while others will provide consumers with foods designed specifically to be healthier and more nutritious.

HEALTH AND NUTRITIONAL BENEFITS

A variety of healthier cooking oils derived from biotechnology are already on the market. Using biotechnology, plant scientists have decreased the total amount of saturated fatty acids in certain vegetable oils. They have also increased the conversion of linoleic acid to the fatty acid found mainly in fish that is associated with lowering cholesterol levels.

Another nutritional concern related to edible oils is the negative health effects produced when vegetable oils are hydrogenated to increase their heat stability for cooking or to solidify oils used in making margarine. The hydrogenation process results in the formation of trans-fatty acids.

Biotechnology companies have given soybean oil these same properties, not through hydrogenation, but by using biotechnology to increase the amount of the naturally occurring fatty acid, stearic acid.

Animal scientists are also using biotechnology to create healthier meat products, such as beef with lower fat content and pigs with a higher meat-to-fat ratio.

Other health and nutritional benefits of crops improved through biotechnology include increased nutritional value of crops, especially those that are food staples in developing countries. Scientists at Nehru University in New Delhi used a gene found in the South American plant amaranth to increase the protein content of potatoes by 30 percent. These transgenic potatoes also contain large amounts of essential amino acids not found in unmodified potatoes. Other examples include golden rice and canola oil, both of which are high in vitamin A. The golden rice developers further improved rice with two other genes that increase the amount and digestibility of iron.

Biotechnology also promises to improve the health benefits of *functional foods*. Functional foods are foods containing significant levels of biologically active components that impart health benefits beyond our basic needs for sufficient calories, essential amino acids, vitamins and minerals. Familiar examples of functional foods include compounds in garlic and onions that lower cholesterol and improve the immune response; antioxidants found in green tea; and the glucosinolates in broccoli and cabbage that stimulate anticancer enzymes.

We are using biotechnology to increase the production of these compounds in functional foods. For example, researchers at Purdue University and the U.S. Department of Agriculture, created a tomato variety that contains three times as much of the antioxidant lycopene as the unmodified variety. Lycopene consumption is associated with a lower risk of prostate and breast cancer and decreased blood levels of "bad cholesterol." Other USDA researchers are using biotechnology to increase the amount of ellagic acid, a cancer protective agent, in strawberries.

PRODUCT QUALITY

We are also using biotechnology to change the characteristics of the raw material inputs so that they are more attractive to consumers and more amenable to processing. Biotechnology researchers are increasing the shelf life of fresh fruits and vegetables; improving the crispness of carrots, peppers and celery; creating seedless varieties of grapes and melons; extending the seasonal geographic availability of tomatoes, strawberries and raspberries; improving the flavor of tomatoes, lettuce, peppers, peas and potatoes; and creating caffeine-free coffee and tea.

Japanese scientists have now identified the enzyme that produces the chemical that makes us cry when we slice an onion. Knowing the identity of the enzyme is the first step in finding a way to block the gene to create "tearless" onions.

Much of the work on improving how well crops endure food processing involves changing the ratio of water to starch. Potatoes with higher starch content are healthier because they absorb less oil when they are fried, for example. Another important benefit is that starchier potatoes require less energy to process and therefore cost less to handle. Many tomato processors now use tomatoes derived from a biotechnology technique, somaclonal variant selection. The new tomatoes, used in soup, ketchup and tomato paste, contain 30 percent less water and are processed with greater efficiency.

Another food processing sector that will benefit economically from better quality raw materials is the dairy products industry. Scientists in New Zealand have now used biotechnology to increase the amount of the protein casein, which is essential to cheese making, in milk by 13 percent.

Biotechnology also allows the economically viable production of valuable, naturally occurring compounds that cannot be manufactured by other means. For example, commercial-scale production of the natural and highly marketable sweetener known as fructans has long eluded food-processing engineers. Fructans, which are short chains of the sugar molecule fructose, taste like sugar but have no calories. Scientists found a gene that converts 90 percent of the sugar found in beets to fructans. Because 40 percent of the transgenic beet dry weight is fructans, this crop can serve as a manufacturing facility for fructans.

SAFETY OF THE RAW MATERIALS

The most significant food-safety issue food producers face is microbial contamination, which can occur at any point from farm to table. Any biotechnology product that decreases microbes found on animal products and crop plants will significantly improve the safety of raw materials entering the food supply. Improved food safety through decreased microbial contamination begins on the farm. Transgenic disease-resistant and insect-resistant crops have less microbial contamination.

Biotechnology is improving the safety of raw materials by helping food scientists discover the exact identity of the allergenic protein in foods such as peanuts, soybeans and milk, so they can then remove them. Although 95 percent of food allergies can be traced to a group of eight foods, in most cases we do not know which of the thousands of proteins in a food triggered the reaction. With biotechnology techniques, we are making great progress in identifying these allergens. More importantly, scientists have succeeded in using biotechnology to block or remove allergenicity genes in peanuts, soybeans and shrimp.

Finally, biotechnology is helping us improve the safety of raw agricultural products by decreasing the amount of natural plant toxins found in foods such as potato and cassava.

Food Processing

Microorganisms have been essential to the food-processing industry for decades. They play a role in the production of the fermented foods listed in Table 1. They also serve as a rich source of food additives, enzymes and other substances used in food processing.

IMPROVING FOOD FERMENTORS

Because of the importance of fermented foods to so many cultures, scientists are conducting a lot of work to improve the microorganisms that carry out food fermentations. The bacterium responsible for many of our fermented dairy products, such as cheese and yogurt, is susceptible to infection by a virus that causes substantial economic losses to the food industry. Through recombinant technology, researchers have made some strains of this bacterium and other important fermentors resistant to viral infection.

We have known for years that some bacteria used in food fermentation produce compounds that kill other, contaminating bacteria that cause food poisoning and food spoilage. Using biotechnology we are equipping many of our microbial fermentors with this self-defense mechanism to decrease microbial contamination of fermented foods.

FOOD ADDITIVES AND PROCESSING AIDS

Microorganisms have been essential to the food industry not only for their importance as fermentors, but also because they are the source of many of the additives and processing aids used in food processing. Biotechnology advances will enhance their value to the food industry even further.

Food additives are substances used to increase nutritional value, retard spoilage, change consistency and enhance flavor. The compounds food processors use as food additives are substances nature has provided and are usually of plant or microbial origin, such as xanthan gum and guar gum, which are produced by microbes. Many of the amino acid supplements, flavors, flavor enhancers and vitamins added to breakfast cereals are produced by microbial fermentation. Through biotechnology, food processors will be able to produce many compounds that could serve as food additives but that now are in scant supply or that are found in microorganisms or plants difficult to maintain in fermentation systems.

Food processors use plant starch as a thickener and fat substitute in low-fat products. Currently, the starch is extracted from plants and modified using chemicals or energy-consuming mechanical processes. Scientists are using biotechnology to change the starch in crop plants so that it no longer requires special handling before it can be used.

Table: Microbial fermentation is essential to the production of these fermented foods

beer	distilled	tamari
bologna	liquors	tea
bread/baked	kefir	tempeh
goods	miso	tofu
buttermilk	olives	vinegar

cheeses	pickles	wine
cider	salami	yogurt
cocoa	sauerkraut	
coffee	sour cream	
cottage cheese	soy sauce	

Enzymes produced by microbial fermentation play essential roles as processing aids in the food industry. The first commercial food product produced by biotechnology was an enzyme used in cheese making. Prior to biotech techniques, this enzyme had to be extracted from the stomach of calves, lambs and baby goats, but it is now produced by microorganisms that were given the gene for this enzyme.

The production of high-fructose corn syrup from cornstarch requires three enzymes, and those same enzymes are important in making baked goods and beer. Other enzymes are essential to the production of fruit juices, candies with soft centres, and cheeses. The food industry uses more than 55 different enzyme products in food processing. This number will increase as we discover how to capitalize on the extraordinary diversity of the microbial world and obtain new enzymes that will prove important in food processing.

Food Safety Testing

In addition to the many ways biotechnology is helping us enhance the safety of the food supply, biotechnology is providing us with many tools to detect microorganisms and the toxins they produce. Monoclonal antibody tests, biosensors, polymerase chain reaction (PCR) methods and DNA probes are being developed that will be used to determine the presence of harmful bacteria that cause food poisoning and food spoilage, such as *Listeria* and *Clostridium botulinum*.

We can now distinguish *E. coli* 0157:H7, the strain of *E. coli* responsible for several deaths in recent years, from the many other harmless E. coli strains. These tests are portable, quicker and more sensitive to low levels of microbial contamination than previous tests because of the increased specificity of molecular technique. For example, the new diagnostic tests for *Salmonella* yield results in 36 hours, compared with the three or four days the older detection methods required.

Biotechnology-based diagnostics have also been developed that allow us to detect toxins, such as aflatoxin, produced by fungi and molds that grow on crops, and to determine whether food products have inadvertently been contaminated with peanuts, a potent allergen.

Chapter 2

Food-based Vitamins and Minerals

Dietary patterns have varied over time depending on the agricultural practices and the climatic, ecologic, cultural, and socio-economic factors, which determine available foods. At present, virtually all dietary patterns adequately satisfy or even exceed the nutritional needs of population groups. This is true except where socio-economic conditions limit the capacity to produce and purchase food or aberrant cultural practices restrict the choice of foods. It is thought that if people have access to a sufficient quantity and variety of foods, they will meet their nutritional needs. The current practice of evaluating nutritive value of diets should include not only energy and protein adequacy but also the micronutrient density of the diet.

A healthy diet can be attained in more than one way because of the variety of foods, which can be combined. It is thus difficult to define the ranges of intake for a specific food, which should be included in a given combination to comply with nutritional adequacy. In practice, the set of food combinations which is compatible with nutritional adequacy is restricted by the level of food production sustainable in a given ecologic and population setting. In addition, there are economic constraints, which limit food supply at household level. The development of food-based dietary guidelines (FBDGs) recognises this and focuses on the combination of foods that can meet nutrient requirements rather than on how each specific nutrient is provided in adequate amounts.

The first step in the process of setting dietary guidelines is defining the significant diet-related public health problems in a community. Once these are defined, the adequacy of the diet is evaluated by comparing the information available on dietary intake with recommended nutrient intakes (RNIs). Nutrient intake goals under this situation are specific for a given ecologic setting, and their purpose is to promote overall health, control specific nutritional diseases (whether they are induced by an excess or deficiency of nutrient intake), and reduce the risk of diet-related multi-factorial diseases.

Dietary guidelines represent the practical way to reach the nutritional goals for a given population. They take into account the customary dietary pattern and indicate what aspects should be modified. They consider the ecologic setting, socio-economic and cultural factors, and biologic and physical environment in which the population lives.

The alternative approach to defining nutritional adequacy of diets is based on the biochemical and physiologic basis of human nutritional requirements in health and disease. The quantitative definition of nutrient needs and its expression as RNIs have been important instruments of food and nutrition policy in many countries and have focused the attention of international bodies.

This nutrient-based approach has served many purposes but has not always fostered the establishment of nutritional and dietary priorities consistent with the broad public health priorities at the national and international levels. It has permitted a more precise definition of requirements for essential nutrients when establishing RNIs but unfortunately has often been narrowly focused, concentrating on the precise nutrient requirement amount and not on solving the nutritional problems of the world. In contrast to RNIs, FBDGs are based on the fact that people eat food, not nutrients. The notion of nutrient density is helpful for defining FBDGs and evaluating the adequacy of diets. In addition, they serve to educate the public through the mass media and provide a practical guide to selecting foods by defining dietary adequacy.

Advice for a healthy diet should provide both a quantitative and qualitative description of the diet for it to be understood by individuals, who should be given information on both sise and number of servings per day. The quantitative aspects include the estimation of the amount of nutrients in foods and their bio-availability in the form they are actually consumed. Unfortunately, available food composition data for most foods currently consumed in the world are incomplete, outdated, or insufficient for evaluating true bio-availability. The qualitative aspects relate to the biologic utilisation of nutrients in the food as consumed by humans and explore the potential for interaction among nutrients. Such an interaction may enhance or inhibit the bio-availability of a nutrient from a given food source.

Including foods in the diet, which have high micronutrient density - such as pulses or legumes, vegetables (including green leafy vegetables), and fruits - is the preferred way of ensuring optimal nutrition including micronutrient adequacy for most population groups. Most population groups afflicted by micronutrient deficiency largely subsist on refined cereal grain or tuber-based diets, which provide energy and protein (with improper amino acid balance) but are insufficient in critical micronutrients.

The recent interest in the role of phyto-chemicals and antioxidants on health and their presence in plant foods lend further support to the recommendation for increasing vegetables and fruit consumed in the diet. The need for dietary diversification is supported by the knowledge of the interrelationships of food components, which may enhance the nutritional value of foods and prevent undesirable imbalances, which may limit the utilisation of some nutrients. For example, fruits rich in ascorbic acid will enhance the absorption of ionic iron.

If energy intake is low, for example, in the case of young children, sedentary women, or the elderly, the diet may not provide vitamin and mineral intakes sufficient to meet the RNIs. This situation may be of special relevance to the elderly, who are inactive, have decreased lean body mass, and typically decrease their energy intake. Young children, pregnant women, and lactating women, who have greater micronutrient needs relative to their energy needs, will also require increased micronutrient density.

The household is the basic unit for food consumption under most settings, and if there is sufficient food, individual members of the household can consume a diet with the recommended nutrient densities and meet their specific RNIs. However, appropriate food distribution within the family must be considered to ensure that children and women receive adequate food with high micronutrient density. Household food distribution must be considered when establishing general dietary guidelines and addressing the needs of vulnerable groups in the community. In addition, education detailing the appropriate storage and processing of foods to prevent micronutrient losses at the household level is important.

DIETARY DIVERSIFICATION WHEN CONSUMING CEREAL AND TUBER-BASED DIETS

Dietary diversification is important to improve the intake of critical nutrients. The micronutrients selected discussed here, although limited in number, are of public health relevance or serve as markers for overall micronutrient intake. The nutrients selected for discussion below include some of the nutrients, which are most difficult to obtain in cereal and tuber-based diets. Nutrient deficiencies of vitamin A, iron, and zinc are widespread.

Vitamin A

The vitamin A content of most staple diets can be significantly improved with the addition of a relatively small portion of plant foods rich in carotenoids, the precursors of vitamin A. For example, a usual portion of cooked carrots (50 g) added to a daily diet, or 21 g of carrots per 4. 184 MJ, provides 500 mg retinol equivalents, which is the recommended nutrient density for this vitamin. The biologic activity of pro-vitamin A varies among different plant sources, and fruits and vegetables such as carrots, mango, papaya, and melon contain large amounts of nutritionally active carotenoids. Green leafy vegetables such as ivy gourd have been successfully used in Thailand as a source of vitamin A, and carotenoid-rich red palm oil serves as an easily available and excellent source of vitamin A in other countries.

Consequently, a regular portion of these foods included in an individual's diet may provide 100 percent or more of the daily requirement for retinol equivalents. Vitamin A is also present in animal food sources in a highly bio-available form. Therefore it is important to consider the possibility of meeting vitamin A needs by including animal foods in the diet. For example, providing minor amounts of fish or chicken liver (20-25 g) in the diet provides more than the recommended vitamin A nutrient density for virtually all age and sex groups.

Vitamin C

A real gain in vitamin C intake can be achieved by including citrus fruit or other foods rich in ascorbic acid in the diet. For example, an orange or a small amount of other vitamin C-rich fruit (60 g of edible portion) provides the recommended ascorbic acid density. Adding an orange to a potato-based diet increases the level of vitamin C threefold. Other good vitamin C food sources are guava, amla, kiwi, cranberries, strawberries, papaya, mango, melon, cantaloupe, spinach, Swiss chard, tomato, asparagus, and Brussels sprouts. All these foods, when added to a diet or meal in regular portion sises, will significantly improve the vitamin C density. Because ascorbic acid is heat labile, minimal cooking (steaming or stir-frying) is recommended to maximise the bio-available nutrient. The significance of consuming vitamin C with meals will be discussed relative to iron absorption.

Folate

Folate is now considered significant not only for the prevention of macrocytic anaemia, but also for normal foetal development. Recently, this vitamin was implicated in the maintenance of cardiovascular health and cognitive function in the elderly. Staple diets consisting largely of cereal grains and tubers are very low in folate but can be improved by the addition of legumes or green leafy vegetables. For example, a regular portion of cooked lentils (95 g) added to a rice-based diet can provide an amount of folate sufficient to meet the desirable nutrient density for this vitamin.

Other legumes such as beans and peas are also good sources of this vitamin, but larger portions are needed for folate sufficiency. Cluster bean and colacasia leaves are excellent folate sources used in the Indian diet. Another good source of folate is chicken liver; only one portion (20-25 g) is sufficient to meet the desirable nutrient density for folate and vitamin A simultaneously. The best sources of folate are organ meats, green leafy vegetables, and sprouts. However, 50 percent or more of food folate is destroyed during cooking. Prolonged heating in large volumes of water should be avoided, and it is advisable to consume the water used in the cooking of vegetables.

Iron and Zinc

Minerals such as iron and zinc are low in cereal and tuber-based diets, but the addition of legumes can slightly improve the iron content of those diets. However, the bio-availability of this non-heme iron source is low. Therefore, it is not possible to meet the recommended levels of iron and zinc in the staple-based diets through a food-based approach unless some meat, poultry, or fish is included. For example adding a small portion (50 g) of meat, poultry, or fish will increase the total iron content as well as the amount of bio-available iron. For zinc the presence of a small portion (50 g) of meat, poultry, or fish will secure dietary sufficiency of most staple diets.

The consumption of ascorbic acid along with the food rich in iron will enhance absorption. There is a critical balance between enhancers and inhibitors of iron absorption. Nutritional status can be improved significantly by educating households on food preparation practices, which minimise the consumption of inhibitors of iron absorption; for example, the fermentation of phytate-containing grains before the baking of breads to enhance iron absorption.

PRACTICES WHICH WILL ENHANCE THE SUCCESS OF FOOD-BASED APPROACHES

To achieve dietary adequacy of vitamin A, vitamin C, folate, iron, and zinc by using food-based approaches, food preparation and dietary practices must be considered. For example, it is important to recommend that vegetables rich in vitamin C, folate, and other water-soluble or heat-labile vitamins be minimally cooked in small amounts of water. For iron bio-availability it is essential to reduce the intake of inhibitors of iron absorption and to increase the intake of enhancers of absorption in a given meal.

Following this strategy, it is recommended to increase the intake of: germinated seeds, fermented cereals, heat-processed cereals, meats, and fruits and vegetables rich in vitamin C and to encourage the consumption of tea, coffee, chocolate, or herbal teas at times other than with meals. Consumption of flesh foods improves zinc absorption whereas it is inhibited by consumption of diets high in phytate, such as diets based on unrefined cereal. Zinc availability can be estimated according to the phytate-to-zinc (molar) ratio of the meal.

This advice is particularly important for people who consume cereal and tuber-based diets. These foods constitute the main staples for most populations of the world, populations that are also most at risk for micronutrient deficiencies. Other alternatives - fortification and supplementation - have been proposed as stopgap measures when food-based approaches are not feasible or are still in progress. There is a definite role for fortification in meeting iron, folate, iodine, and zinc needs.

Fortification and supplementation should be seen as complementary to food-based strategies and not as a replacement. Combined, all these strategies can go a long way toward stabilising the micronutrient status of populations at risk. Food-based approaches usually take longer to implement but once established are truly sustainable.

DELINEATING THE ROLE OF SUPPLEMENTATION AND FOOD FORTIFICATION FOR NUTRIENTS WHICH CANNOT BE SUPPLIED BY REGULAR FOODS

Under ideal conditions of food access and availability, food diversity should satisfy micronutrient and energy needs of the general population. Unfortunately, for many people in the world, the access to a variety of micronutrient-rich foods is not possible. As demonstrated in our analysis of cereal and tuber-based diets, micronutrient-rich foods including small amount of flesh foods and a variety of plant foods are needed daily. This may not be realistic at present for many communities living under conditions of poverty. Food fortification and food supplementation are important alternatives that complement food-based approaches to satisfy the nutritional needs of people in developing and developed countries.

Fortification

Fortification refers to the addition of nutrients to a commonly eaten food. It is possible for a single nutrient or group of micronutrients to be added to the vehicle, which has been identified through a process in which all stakeholders have participated. This strategy is accepted as sustainable under most conditions and often is cost effective on a large scale when successfully implemented. Iron fortification of wheat flour and iodine fortification of salt is examples of fortification strategies with excellent results.

There are at least three essential conditions that must be met in any fortification programme: the fortificant should be effective, bio-available, acceptable, and affordable; the selected food vehicle should be easily accessible and a specified amount of it should be regularly consumed in the local diet; and detailed production instructions and monitoring procedures should be in place and enforced by law.

Iron fortification

Food fortification with iron is recommended when dietary iron is insufficient or the dietary iron is of poor bio-availability, which is the reality for most people in the developing world and for vulnerable population groups in the developed world. Moreover, the prevalence of iron deficiency and anaemia in vegetarians and in populations of the developing world which rely on cereal or tuber foods is significantly higher than in omnivore populations.

Iron is present in foods in two forms, as heme iron, which is derived from flesh foods (meats, poultry, and fish), and as non-heme iron, which is the inorganic form present in plant foods such as legumes, grains, nuts, and vegetables. Heme iron is highly (20-30 percent) absorbed and its bio-availability is relatively unaffected by dietary factors. Non-heme iron has a lower rate of absorption (2-10 percent), depending on the balance between iron absorption inhibitors (phytates, polyphenols, calcium, and phosphate) and iron absorption enhancers (ascorbic and citric acids, cysteine-containing peptides, ethanol, and fermentation products) present in the diet. Because staple foods around the world provide predominantly non-heme iron sources of low bio-availability, the traditionally eaten staple foods represent an excellent vehicle for iron fortification. Examples of foods, which have been fortified, are wheat flour, corn (maise) flour, rice, salt, sugar, cookies, curry powder, fish sauce, and soy sauce. Nevertheless the beneficial effects of consumption of iron absorption enhancers have been extensively proven and should always be promoted.

Iodine Fortification

Iodine is sparsely distributed in the Earth's surface and foods grown in soils with little or no iodine lack an adequate amount of this micronutrient. This situation had made iodine deficiency disorders exceedingly common in most of the world and highly prevalent in many countries before the introduction of salt iodisation. Only foods of marine origin are naturally rich sources of iodine. Salt is a common food used by most people worldwide, and the establishment of an well-implemented permanent salt-iodisation programme has been proven to eradicate iodine deficiency disorders.

However, salt iodisation is not simply a matter of legislating mandatory iodisation of salt. It is important to determine the best fortification technique, co-ordinate the implementation at all salt production sites, establish effective monitoring and quality control programmes, and measure iodine fortification level periodically. The difficulties in implementing salt iodisation programmes arise primarily when the salt industry is widely dispersed among many small producers. The level of iodine fortification usually lies between 25 and 50 mg/kg salt.

The actual amount should be specified according to the level of salt intake and magnitude of deficit at the country level, because iodine must be added within safe and effective ranges. Additionally, it is very important to implement a monitoring plan to control the amount of iodine in the salt at the consumer's table. United Nations agencies responsible for assisting governments in establishing iodisation programmes should provide technical support for programme implementation, monitoring, and evaluation to ensure sustainability.

Zinc Fortification

The body depends on a regular zinc supply provided by the daily diet because stores are quite limited. Food diversity analysis demonstrates that it is virtually impossible to achieve zinc adequacy in the absence of a flesh food source. Among flesh foods, beef is the best source of zinc and is followed by poultry and then fish. Zinc fortification programmes are being studied, especially for populations, which consume predominately plant foods. Fortification of cereal staple foods is a potentially attractive intervention, which could benefit the whole population as well as target the vulnerable population groups of children and pregnant women. Such addition of zinc to the diet would perhaps decrease the prevalence of stunting in many developing countries with low-zinc diets, because linear growth is affected by zinc supply.

Folic Acid Fortification

The recommended nutrient density by the developers of the FAO/WHO FBDGs for folic acid is 200 mg/4. 184 MJ. Although this reference value is higher than other standards of reference, the increase in folic acid consumption by women of childbearing age is very important: it may improve birth weight and reduce the prevalence of neural tube defects by 50 percent. Elevated plasma homo-cysteine levels are considered to be an independent risk factor for heart disease; a higher intake of folic acid may also benefit the rest of the population because it may lower homo-cysteine levels in adults. In addition, folate may improve the mental condition of the elderly population.

Although the desirable folic acid density may be achieved through dietary diversity, it requires the daily presence of organ meats, green leafy vegetables, pulses, legumes, or nuts in the diet. Most population groups may not easily reach the appropriate level of folic acid consumption; therefore, folic acid fortification has been recommended. The United States initiated mandatory folic acid fortification of cereal-grain products. The fortification level approved in the United States is 140 mg/100 g product, which will increase the average woman's intake by only 100 mg/day. This amount is considered safe (a dose, which will not mask pernicious anaemia, which results from vitamin B12 deficiency,) but it may be ineffective in lowering the occurrence of neural tube defects.

Supplementation

Supplementation refers to periodic administration of pharmacologic preparations of nutrients as capsules or tablets or by injection when substantial or immediate benefits are necessary for the group at risk. Nutritional supplementation should be restricted to vulnerable groups, which cannot meet their nutrient needs through food (women of childbearing age, infants and young children, elderly people, low socio-economic groups, displaced people, refugees, and populations experiencing other emergency situations). For example, iron supplementation is recognised as the only option to control or prevent iron deficiency anaemia in pregnant women. Supplementation with folic acid should be considered for women of childbearing age who have had a child with neural tube defect to prevent recurrence.

Food-based Dietary Guidelines

Food-based dietary guidelines (FBDGs) are an instrument of and expression of food and nutrition policy and should be based directly on diet and disease relationships of particular relevance to the individual country. Their primary purpose is to educate healthcare professionals and consumers about health promotion and disease prevention. In this way priorities in establishing dietary guidelines can address the relevant public health concerns whether they are related to dietary insufficiency or excess. In this context, meeting the nutritional needs of the population takes its place as one of the components of food and nutrition policy goals along with the priorities included in the FBDGs for improved health and nutrition for a given population.

The world nutrition and health situation demonstrates that the major causes of death and disability have been traditionally related to undernutrition in developing countries and to the imbalance between energy intake and expenditure (which lead to obesity and other chronic diseases - diabetes, cardiovascular disease, hypertension, and stroke) in industrialised countries.

The tragedy is that many suffer from too little food while others have diseases resulting from too much food, but both would benefit from a more balanced distribution of food and other resources. Although the nature of the health and nutrition problems in these two contrasting groups is very different, the dietary guidelines required to improve both situations are not. Most countries presently have the combined burden of malnutrition from deficit and increasing prevalence of obesity and other chronic diseases from over consumption. The approaches to address the problems, nevertheless, should be country and population specific.

Although two-thirds of the world's population depends on cereal or tuber-based diets, the other one-third consumes significant amounts of animal food products. The latter group places an undue demand on land, water, and other resources required for intensive food production, which makes the typical Western diet not only undesirable from the standpoint of health but also environmentally unsustainable. If we balance energy intake with the expenditure required for basal metabolism, physical activity, growth, and repair, we will find that the dietary quality required for health is essentially the same across population groups.

Efforts in nutrition education and health promotion should include a strong encouragement for active lifestyles. Improving energy balance for rural populations in developing countries may mean increasing energy intake to normalise low body mass index (BMI, weight/height^2, calculated as kg/m^2), ensuring adequate energy stores and energy for appropriate social interactions. In sedentary urban populations, improving energy balance will mean increasing physical activity to decrease energy stores (body fat mass) and thus normalise BMI.

Thus, the apparent conflicting goals - eradicating undernutrition while preventing overnutrition - are resolved by promoting sufficient energy for a normal BMI. Moreover, if we accept that FBDGs should be ecologically sustainable, the types and amounts of foods included in a balanced diet are not very different for promoting adequate nutrition in the undernourished and preventing overnutrition in the affluent.

This is well exemplified by the similarities in the FBDGs across countries, whether represented by pyramids, rainbows, dishes, pots, etc. It is obvious that consumption of excess energy will induce an increase in energy stores, which may lead to obesity and related health complications. Populations should consume nutritionally adequate and varied diets, based primarily on foods of plant origin with small amounts of added flesh foods. Households should select predominantly plant-based diets rich in a variety of vegetables and fruits, pulses or legumes, and minimally processed starchy staple foods. The evidence that such diets will prevent or delay a significant proportion of non-communicable chronic diseases is consistent.

A predominantly plant-based diet has a low energy density, which may protect against obesity. This should not exclude small amounts of animal foods, which may make an important nutritional contribution to plant-food-based diets, as illustrated in the examples presented earlier. Inadequate diets occur when food is scarce or when food traditions change rapidly, as is seen in societies undergoing demographic transitions or rapid urbanisation. Traditional diets, when adequate and varied, are likely to be generally healthful and more protective against chronic non-communicable diseases than the typical Western diet, consumed predominantly in industrialised societies.

Reorienting food production, agricultural research, and commercialisation policies needs to take into consideration FBDGs, which increase the demand for a variety of micronutrient-rich foods and thus stimulate production to meet the consumption needs. Prevailing agricultural policies encourage research on and production and importation of foods, which do not necessarily meet the requirements of FBDG implementation. For example, great emphasis is placed on cereals, horticultural crops for export, legumes for export, non-food cash crops, and large livestock.

Necessary policy reorientation is required to ensure increased availability of micronutrient-rich foods within the local food system. Norway has successfully implemented agricultural and food production policies based on a National Nutrition Plan of Action, providing economic incentives for the producer and consumer in support of healthful diets. The results speak for themselves, as Norway has experienced a sustained improvement in life expectancy and a reduction in deaths from cardiovascular disease and other chronic non-communicable conditions.

Chapter 3

Vitamin B-complex

The B-complex vitamins covered here are presented in Table along with the biochemical and physiologic roles of the co-enzyme forms and a brief description of clinical deficiency symptoms.

Table: Physiologic roles and deficiency signs of B-complex vitamins

Vitamin	Physiologic roles	Deficiency
Thiamin (B_1)	Co-enzyme functions in metabolism of carbohydrates and branched-chain amino acids	Beri-beri, polyneuritis, and Wernicke-Korsakoff syndrome
Riboflavin (B_2)	Co-enzyme functions in numerous oxidation and reduction reactions	Growth, cheilosis, angular stomatitis, and dermatitis
Niacin (nicotinic acid and nicotinamide)	Co-substrate/co-enzyme for hydrogen transfer with numerous dehydrogenases	Pellagra with diarrhoea, dermatitis, and dementia
Vitamin B_6 (pyridoxine, pyridoxamine, and pyridoxal)	Co-enzyme functions in metabolism of amino acids, glycogen, and sphingoid bases	Naso-lateral seborrhoea, glossitis, and peripheral neuropathy (epileptiform convulsions in infants)
Pantothenic acid	Constituent of co-enzyme A and phosphopantetheine involved in fatty acid metabolism	Fatigue, sleep disturbances, impaired coordination, and nausea
Biotin	Co-enzyme functions in bicarbonate-dependent carboxylations	Fatigue, depression, nausea, dermatitis, and muscular pains

Rice and wheat are the staples for many populations of the world. Excessive refining and polishing of cereals removes considerable proportions of B vitamins contained in these cereals. Clinical manifestations of deficiency of some B vitamins - such as beri-beri (cardiac and dry), peripheral neuropathies, pellagra, and oral and genital lesions (related to riboflavin deficiency) - were once major public health problems in parts of the world. These manifestations have now declined, the decline being brought about not through programmemes, which distribute synthetic vitamins but through changes in the patterns of food availability and consequent changes in dietary practices of the populations.

Although these clinical manifestations of B-vitamin deficiencies have decreased, there is evidence of widespread sub-clinical deficiency of these vitamins (especially of riboflavin and pyridoxine). These sub-clinical deficiencies, although less dramatic in their manifestations, exert deleterious metabolic effects. Despite the progress in reduction of large-scale deficiency in the world, there are periodic reports of outbreaks of B-complex deficiencies, which are linked to deficits of B vitamins in populations under various distress conditions.

Refugee and displaced population groups are at risk for B-complex deficiency because most cereal foods used under emergency situations are not fortified with micronutrients. Recent reports have implicated the low B-complex content of diets as a factor in the outbreak of peripheral neuropathy and visual loss observed the adult population of Cuba. This deficiency in Cuba resulted from the consequences of an economic blockade.

THIAMIN

Deficiency

Thiamin (vitamin B_1, aneurin) deficiency results in the disease called beri-beri, which has been classically considered to exist in dry (paralytic) and wet (oedematous). Beri-beri occurs in human-milk-fed infants whose nursing mothers are deficient. It also occurs in adults with high carbohydrate intakes mainly from milled rice and with intakes of anti-thiamin factors. Beri-beri is still endemic in Asia. In relatively industrialised nations, the neurologic reflections of Wernicke-Korsakoff syndrome are frequently associated with chronic alcoholism with limited food consumption. Some cases of thiamin deficiency have been observed with patients who are hypermetabolic, are on parenteral nutrition, are undergoing chronic renal dialysis, or have undergone a gastrectomy.

Toxicity

Toxicity is not a problem with thiamin because renal clearance of levels conceivably ingested is rapid.

Functions

Thiamin functions as the co-enzyme thiamin pyrophosphate (TPP) in the metabolism of carbohydrates and branched-chain amino acids. Specifically the Mg^{2+}-coordinated TPP participates in the formation of a-ketols (e. g., among hexose and pentose phosphates) as catalysed by transketolase and in the oxidation of a-keto acids (e. g., pyruvate, a-ketoglutarate, and branched-chain a-keto acids) by dehydrogenase complexes. Hence, when there is insufficient thiamin, the overall decrease in carbohydrate metabolism and its inter-connection with amino acid metabolism (via a-keto acids) have severe consequences, such as a decrease in the formation of acetylcholine for neural function.

Biochemical Indicators

Indicators used to estimate thiamin requirements are urinary excretion, erythrocyte transketolase activity coefficient, erythrocyte thiamin, blood pyruvate and lactate, and neurologic changes. The excretion rate of the vitamin and its metabolites reflects intake, and the validity of the assessment of thiamin nutriture is improved with load test. Erythrocyte transketolase activity co-efficient reflects TPP levels and can indicate rare genetic defects. Erythrocyte thiamin is mainly a direct measure of TPP but also is a measure of thiamin and thiamin monophosphate by high performance liquid chromatography (HPLC) separation.

Thiamin status has been assessed by measuring urinary thiamin excretion under basal conditions or after thiamin loading, transketolase activity, and free and phosphorylated forms in blood or serum (6, 9). Although overlap with baseline values for urinary thiamin was found with oral doses below 1 mg, a correlation of 0. 86 between oral and excreted amounts. The erythrocyte transketolase assay, in which an activity coefficient based on a TPP stimulation of the basal level is given, continues to be a main functional indicator, but some problems have been encountered. Gans and Harper found a wide range of TPP effect when thiamin intakes were adequate above 1. 5 mg/day over a 3-day period.

In some cases the activity coefficient may appear normal after prolonged deficiency. This measure seemed poorly correlated with dietary intakes estimated for a group of English adolescents. Certainly, there are both inter-individual and genetic factors affecting the transketolase. Baines and Davies suggested that it is useful to determine erythrocyte TPP directly because the co-enzyme is less susceptible to factors that influence enzyme activity; however, there are also methods for determining thiamin and its phosphate esters in whole blood.

Findings by Age and Life Stage

Recommendations for infants are based on adequate food intake. Mean thiamin content of human milk is 0. 21 mg/l (0. 62 mmol/l), which corresponds to 0. 16 mg (0. 49mmol) thiamin per 0. 75 l of secreted milk per day. The blood concentration for total thiamin averages 210 ± 53 nmol/l for infants up to 6 months but decreases over the first 12-18 months of life.

Table: Recommended nutrient intakes for thiamin

	Group mg/day	Recommended nutrient intake
Infants and children		
0-6 months	0. 2	
7-12 months	0. 3	
1-3 years	0. 5	

4-6 years	0. 6
7-9 years	0. 9
dolescents, 10-18 years	
Females	1. 1
Males	1. 2
Adults	
Females, 19+ years	1. 1
Males, 19+ years	1. 2
Pregnancy	1. 4
Lactation	1. 5

RIBOFLAVIN

Deficiency

Riboflavin (vitamin B_2) deficiency results in the condition of hypo- or ariboflavinosis, with sore throat; hyperaemia; oedema of the pharyngeal and oral mucous membranes; cheilosis; angular stomatitis; glossitis; seborrheic dermatitis; and normochromic, normocytic bone marrow. Because the deficiency almost invariably occurs combined with a deficiency of other B-complex vitamins, some of the symptoms (e. g., glossitis and dermatitis) may result from other complicating deficiencies.

The major cause of hypo-riboflavinosis is inadequate dietary intake as a result of limited food supply, which is sometimes exacerbated by poor food storage or processing. Children in developing countries will commonly demonstrate clinical signs of riboflavin deficiency during periods of the year when gastrointestinal infections are prevalent. Decreased assimilation of riboflavin also results from abnormal digestion such as that which occurs with lactose intolerance. This condition is highest in African and Asian populations and can lead to a decreased intake of milk as well as an abnormal absorption of the vitamin. Absorption of riboflavin is also affected in some other conditions, for example, tropical sprue, celiac disease, malignancy and resection of the small bowel, and decreased gastrointestinal passage time. In relatively rare cases the causes of deficiency are inborn errors in which the genetic defect is in the formation of a flavoprotein (e. g., acyl-co-enzyme A [co-A] dehydrogenases). Also at risk are those receiving phototherapy for neonatal jaundice and perhaps those with inadequate thyroid hormone. Some cases of riboflavin deficiency were also observed in Russian school children and Southeast Asian school children.

Toxicity

Riboflavin toxicity is not a problem because of limited intestinal absorption.

Functions

Conversion of riboflavin to flavin mononucleotide (FMN) and further to the predominant flavin adenine dinucleotide (FAD) occurs before these flavins form complexes with numerous flavoprotein dehydrogenases and oxidases. These flavoco-enzymes (FMN and FASD) participate in oxidation-reduction reactions in metabolic pathways and in energy production via the respiratory chain.

Biochemical Indicators

Indicators used to estimate riboflavin requirements are urinary flavin excretion, erythrocyte glutathione reductase activity coefficient, and erythrocyte flavin. The urinary flavin excretion rate of vitamin and metabolites reflects intake; validity of assessment of riboflavin adequacy is improved with load test. Erythrocyte glutathione reductase activity coefficient reflects FAD levels; results are confounded by such genetic defects as glucose-6-phosphate dehydrogenase deficiency and heterozygous b thalassemia.

Erythrocyte flavin is mainly a measure of FMN and riboflavin after hydrolysis of labile FAD and HPLC separation.

Riboflavin status has been assessed by measuring urinary excretion of the vitamin in fasting, random, and 24-hour specimens or by load returns tests (amounts measured after a specific amount of riboflavin is given orally); erythrocyte glutathione reductase; or erythrocyte flavin concentration. The HPLC method with fluorometry gives lower values for urinary riboflavin than do fluorometric methods, which measure the additive fluorescence of similar flavin metabolites.

The metabolites can comprise as much as one-third of total urinary flavin and in some cases may depress assays dependent on a biologic response because certain catabolites can inhibit cellular uptake. Under conditions of adequate riboflavin uptake (»1. 5 mg/day) by adults, an estimated 120 µg (320 nmol) total riboflavin or 80 ìg/g of creatinine is excreted daily.

The erythrocyte glutathione reductase assay, with an activity coefficient (AC) expressing the ratio of activities in the presence and absence of added FAD, continues to be used as a main functional indicator, but some limits have been noted. The reductase in erythrocytes from individuals with glucose-6-phosphate dehydrogenase deficiency (often present in blacks) has an increased avidity for FAD, which makes this test invalid. An upper limit of normality for the AC at 1. 34 based on the mean value plus 2 standard deviations from several hundreds of apparently healthy individuals aged 60 years and over. Suggested guidelines for the interpretation of such enzyme ACs are as follows: less than 1. 2, acceptable; 1. 2-1. 4, low; greater than 1. 4, deficient. In general agreement with earlier findings on erythrocyte flavin, a correlation between cord blood and maternal erythrocyte deficiencies and suggested that values greater than 40 nmol/l are considered adequate.

Factors Affecting Requirements

Several studies reported modest effects of physical activity on the erythrocyte glutathione reductase AC. A slight increase in the AC and decrease in urinary flavin of weight-reducing women and older women undergoing exercise training were "normalised" with 20 percent additional riboflavin. However, riboflavin supplementation did not lead to an increase in work performance when such subjects were not clinically deficient.

Bio-availability of riboflavin in foods, mostly as digestible flavoco-enzymes, is excellent at nearly 95 percent, but absorption of the free vitamin is limited to about 27 mg per single meal or dose in an adult. No more than about 7 percent of food flavin is found as 8a-FAD covalently attached to certain flavoprotein enzymes. Although some portions of the 8a-(amino acid)-riboflavins are released by proteolysis of these flavoproteins, they do not have vitamin activity. A lower fat-to-carbohydrate ratio may decrease the riboflavin requirements of the elderly. Riboflavin interrelates with other B vitamins, notably niacin, which requires FAD for its formation from tryptophan, and vitamin B_6, which requires FMN for conversion of the phosphates of pyridoxine and pyridoxamine to the co-enzyme pyridoxal 5'-phosphate (PLP). Contrary to earlier reports, no difference was seen in riboflavin status of women taking oral contraceptives when dietary intake was controlled by providing a single basic daily menu and meal pattern after 0. 6 mg riboflavin/418 kJ was given in a 2-week acclimation period.

FINDINGS BY AGE AND LIFE STAGE

As reviewed early estimates of riboflavin content in human milk showed changes during the post-partum period. More recent investigations of flavin composition of both human and cow milk have helped clarify the nature of the flavins present and provide better estimates of riboflavin equivalence. For human milk consumed by infants up to age 6 months, the riboflavin equivalence averages 0. 35mg (931 nmol) /l (6) or 0. 26 mg (691nmol) /0. 75 l of milk per day. For low-income Indian women with erythrocyte glutathione reductase activity ratios averaging 1. 80 and a milk riboflavin content of 0. 22 mg/l, breast-fed infants averaged AC ratios near 1. 36 (54). Hence, a deficiency sufficient to reduce human-milk riboflavin content by one-third can lead to a mild subclinical deficiency in infants.

Studies of riboflavin status in adults include those in modestly obese young women on low-energy diets, on deficient Gambians. Most of a 1. 7-mg dose of riboflavin given to healthy adults consuming at least this amount was largely excreted in the urine. Such findings corroborate earlier work indicating a relative saturation of tissue with intakes above 1. 1 mg/day. Studies by Alexander et al. (57) on riboflavin status in the elderly show that doubling the estimated riboflavin intakes of 1. 7 mg/day for women aged 70 years and over, with a reductase AC of 1. 8, led to a doubling of urinary riboflavin from 1. 6-3. 4 μg (4. 2 to 9. 0 nmol) /mg creatinine and a decrease in AC to 1. 25. Pregnant women have an increased erythrocyte glutathione reductase AC.

That riboflavin at 0. 7 mg /4184 kJ was needed to lower the AC of four of eight pregnant women to 1. 3 within 20 days, whereas only 0. 41 mg/4184 kJ was needed for five of the seven non-pregnant women. Maternal riboflavin intake was positively associated with foetal growth in a study of 372 pregnant women. The additional riboflavin requirement of 0. 3 mg/day for pregnancy is an estimate based on increased growth in maternal and foetal compartments. For lactating women, an estimated 0. 3 mg riboflavin is transferred in milk daily and, because utilisation for milk production is assumed to be 70 percent efficient, the value is adjusted upward to 0. 4 mg/day.

Table: Recommended nutrient intakes for riboflavin

Group	Recommended nutrient intake mg/day
Infants and children	
0-6 months	0. 3
7-12 months	0. 4
1-3 years	0. 5
4-6 years	0. 6
7-9 years	0. 9
Adolescents, 10-18 years	
Females	1. 0

Males	1. 3
Adults	
Females, 19+ years	1. 1
Males, 19+ years	1. 3
Pregnancy	1. 4
Lactation	1. 6

NIACIN

Deficiency

Niacin (nicotinic acid) deficiency classically results in pellagra, which is a chronic wasting disease associated with a characteristic erythematous dermatitis that is bilateral and symmetrical, a dementia after mental changes including insomnia and apathy preceding an overt encephalopathy, and diarrhoea resulting from inflammation of the intestinal mucous surfaces. At present, pellagra occurs endemically in poorer areas of India, China, and Africa.

Its cause has been mainly attributed to a deficiency of niacin; however, its biochemical inter-relationship to riboflavin and vitamin B_6, which are needed for the conversion of L-tryptophan to niacin equivalents, suggests that insufficiencies of these vitamins may also contribute to pellagra. Pellagra-like syndromes occurring in the absence of a dietary niacin deficiency are also attributable to disturbances in tryptophan metabolism (e. g., Hartnup disease with impaired absorption of the amino acid and carcinoid syndrome where the major catabolic pathway routes to 5-hydroxytryptophan). Pellagra also occurs in people with chronic alcoholism. Cases of niacin deficiency have been found in people suffering from Crohn's disease.

Functions

Niacin is chemically synonymous with nicotinic acid although the term is also used for its amide (nicotinamide). Nicotinamide is the other form of the vitamin, which does not have the pharmacologic action of the acid that is administered at high doses to lower blood lipids. It is the amide form that exists within the redox-active co-enzymes, nicotinamide adenine dinucleotide (NAD) and its phosphate (NADP), which function in dehydrogenase-reductase systems requiring transfer of a hydride ion.

NAD is also required for non-redox adenosine diphosphate-ribose transfer reactions involved in DNA repair and calcium mobilisation. NAD functions in intracellular respiration and with enzymes involved in the oxidation of fuel substrates such as glyceraldehyde 3-phosphate, lactate, alcohol, 3-hydroxybutyrate, and pyruvate. NADP functions in reductive biosyntheses such as fatty acid and steroid syntheses and in the oxidation of glucose-6-phosphate to ribose-5-phosphate in the pentose phosphate pathway.

Biochemical Indicators

Indicators used to estimate niacin requirements are urinary excretion, plasma concentrations of metabolites, and erythrocyte pyridine nucleotides. The excretion rate of metabolites, mainly N′-methyl-nicotinamide and its 2- and 4-pyridones, reflects intake and is usually expressed as a ratio of the pyridones to N′-methyl-nicotinamide. Concentrations of metabolites, especially 2-pyridone, are measured in plasma after a load test. Erythrocyte pyridine nucleotides measure NAD concentration changes.

Niacin status has been monitored by daily urinary excretion of methylated metabolites, especially the ratio of the 2-pyridone to N′-methyl-nicotinamide; erythrocyte pyridine nucleotides; oral dose uptake tests; erythrocyte NAD; and plasma 2-pyridone. Shibata and Matsuo found that the ratio of urinary 2-pyridone to N′-methyl-nicotinamide was as much a measure of protein adequacy as it was a measure of niacin status. This ratio too insensitive to marginal niacin intake. The ratio of the 2-pyridone to N′-methyl-nicotinamide also appears to be associated with the clinical symptoms of pellagra, principally the dermatitic condition.

In plasma, 2-pyridone levels change in reasonable proportion to niacin intake. Similarly to the situation for erythrocyte pyridine nucleotide (nicotinamide co-enzymes), NAD concentration decreased 70 percent whereas NADP remained unchanged in adult males fed diets with only 6 or 10 mg NEs/day. Erythrocyte NAD provided a marker at least as sensitive as urinary metabolites of niacin in this study and in a niacin depletion study of elderly subjects.

Factors Affecting Requirements

The biosynthesis of niacin derivatives on the pathway to nicotinamide co-enzymes stems from tryptophan, an essential amino acid found in protein, and as such this source of NEs increases niacin intake. There are several dietary, drug, and disease factors that reduce the conversion of tryptophan to niacin. Although a 60-to-1 conversion factor represents the average for human utilisation of tryptophan as NEs, there are substantial individual differences.

There is also an interdependence of enzymes within the tryptophan-to-niacin pathway where vitamin B_6 and riboflavin are functional. Further, riboflavin (as FMN) is required for the oxidase that forms co-enzymic PLP from the alcohol and amine forms of phosphorylated vitamin B_6.

Findings by Age and Life Stage

Niacin content of human milk is approximately 1. 5 mg (12. 3 mmol) /l and the tryptophan content is 210 mg (1. 0mmol) /l. Hence, the total content is approximately 5 mg NEs/l or 4 mg NEs/ 0. 75 l secreted daily in human milk. Recent studies together with those reported in the 1950s suggest that 12. 5 mg NEs, which corresponds to 5. 6 mg NEs/4184 kJ, is minimally sufficient for niacin intake in adults.

For pregnant women, where 230 MJ is the estimated energy cost of pregnancy, calculated needs above those of non-pregnant women are 5. 6 mg NEs/ 4186 kjoule (1,000 kcal) ′ 230,000 kjoule (55,000 kcal), or 308 mg NEs for the entire pregnancy or 1. 7 mg NEs/day (308 mg NEs/180 days) for the second and third trimester, which is about a 10 percent increase. Also about 2 mg NEs/day is required for growth in maternal and foetal compartments.

For lactating women, an estimated 1. 4 mg preformed niacin is secreted daily, and an additional requirement of less than 1 mg is needed to support the energy expenditure of lactation. Hence, 2. 4 mg NEs/day is the added need attributable to lactation.

Table: Recommended nutrient intakes for niacin

Group	Recommended nutrient intake, NEs/day
Infants and children	
0-6 months	2[b]
7-12 months	4
1-3 years	6
4-6 years	8
7-9 years	12
Adolescents, 10-18 years	16
Adults	
Females, 19+ years	14
Males, 19+ years	16
Pregnancy	18
Lactation	17

[a] NEs, niacin equivalents. [b] Preformed.

VITAMIN B_6

Deficiency

A deficiency of vitamin B_6 alone is uncommon because it usually occurs in association with a deficit in other B-complex vitamins. Early biochemical changes include decreased levels of plasma PLP and urinary 4-pyridoxic acid. These are followed by decreases in synthesis of transaminases (aminotransferases) and other enzymes of amino acid metabolism such that there is an increased urinary xanthurenate and a decreased glutamate conversion to the anti-neurotransmitter gaminobutyrate. Hypovitaminosis B_6 may often occur with riboflavin deficiency, because riboflavin is needed for the formation of the co-enzyme PLP. Infants are especially susceptible to insufficient intakes, which can lead to epileptiform convulsions. Skin changes include dermatitis with cheilosis and glossitis. There is usually a decrease in circulating lymphocytes and possibly a normocytic, microcytic, or sideroblastic anaemia. The sensitivity of such systems as sulphur amino acid metabolism to vitamin B_6 availability is reflected in homo-cysteinemia. A decrease in the metabolism of glutamate in the brain, which is found in vitamin B_6 insufficiency, reflects a nervous system dysfunction. As is the case with other micronutrient deficiencies, vitamin B_6 deficiency results in an impairment of the immune system. A current concern is for the rather pandemic occurrence of somewhat low vitamin B_6 intakes in many people who eat poorly. Vitamin B_6 deficiency has also been observed in Russian schoolchildren, Southeast Asian schoolchildren (infected with hookworm), elderly Europeans (Dutch), and in some individuals with hyperhomo-cysteinemia or on chronic hemodialysis. Several medical conditions can also affect vitamin B_6 metabolism and lead to deficiency symptoms.

Toxicity

Use of high doses of pyridoxine for dubious treatment of pre-menstrual syndrome, carpal tunnel syndrome, and some neurologic diseases has resulted in neurotoxicity. A UL of 100 mg/day as proposed by the US Food and Nutrition Board was adopted by this consultation.

Functions

There are three natural vitamers (different forms of the vitamin) of vitamin B_6, namely pyridoxine, pyridoxamine, and pyridoxal. These must be phosphorylated and the 5'-phosphates of the first two oxidised to the functional PLP, which serves as a carbonyl-reactive co-enzyme to diverse enzymes involved in the metabolism of amino acids. Such enzymes include aminotransferases, decarboxylases, and dehydratases, dd-aminolevulinate synthase in heme biosynthesis, phosphorylase in glycogen breakdown, and sphingoid base biosynthesis, etc.

Biochemical Indicators

Indicators used to estimate vitamin B_6 requirements are PLP, urinary excretion, erythrocyte aminotransferases activity coefficients, tryptophan catabolites, erythrocyte and whole blood PLP, and plasma homo-cysteine. PLP is the major vitamin B_6 form in tissue and reflects liver PLP; it changes fairly slowly in response to vitamin intake. The excretion rate of vitamin and particularly 4-pyridoxate reflects intake. Erythrocyte aminotransferases for aspartate and alanine reflect PLP levels and show large variations in activity coefficients. The urinary excretion of xanthurenate, a tryptophan catabolite, is used especially after a tryptophan load test.

Vitamin B_6 status is most appropriately evaluated by using a combination of indicators, including those considered direct (e. g., vitamer concentration in cells or fluids) and indirect or functional indicators (e. g., erythrocyte aminotransferase saturation by PLP or tryptophan metabolites). Plasma PLP may be the best single indicator because it appears to reflect tissue stores. Diets containing less than 0. 05 mg vitamin B_6given to 11 young women led to abnormal electroence phalograph patterns in 2 of the women and a plasma PLP of approximately 9 nmol. Hence, a level of about 10 nmol is considered sub-optimal. A cut-off for plasma PLP of 20 nmol has been proposed as an index of adequacy based on recent findings. Plasma PLP levels have been reported to fall with age.

Urinary 4-pyridoxic acid level changes promptly with changes in vitamin B_6 intake and is therefore of questionable value in assessing status. However, a value higher than 3 μmol/day, achieved with an intake of ~1 mg/d, has been suggested to reflect adequate intake. Erythrocyte aminotransferases for aspartate and alanine are commonly measured before and after addition of PLP to ascertain amounts of apoenzymes, the proportion of which increases with vitamin B_6 depletion. Values of 1. 5-1. 6 for the aspartate aminotransferase and approximately 1. 2 for the alanine aminotransferase have been suggested as being adequate. Catabolites from tryptophan and methionine have also been used to assess vitamin B_6 status. In a review of such literature, Leklem suggested that a 24-hour urinary excretion less than 65 μmol xanthurenate after a 2-g oral dose of tryptophan indicates normal vitamin B_6 status.

Factors Affecting Requirements

A recent review by Gregory shows that bio-availability of vitamin B_6 in a mixed diet is about 75 percent, with approximately 8 percent of this total contributed by pyridoxine b-D-glucoside, which is about half as effectively utilised as free B vitamers or their phosphates. The amine and aldehyde forms of vitamin B_6 may be about 10 percent less effective than pyridoxine. Despite the involvement of PLP with many enzymes affecting amino acid metabolism, there seems to be only a slight effect of dietary proteins on vitamin B_6 status. Studies reported decreases in indicators of vitamin B_6 status in women receiving oral contraceptives, but this probably reflects hormonal stimulation of tryptophan catabolism rather than any deficiency of vitamin B_6 per se. Subjects with pre-eclampsia or eclampsia have lowered plasma PLP levels than do healthy pregnant women.

Findings by Age and Life Stage

The average intake for infants, based on human-milk content, is 0. 13 mg/l or 0. 1 mg/0. 75 l/day. With an average maternal dietary intake of vitamin B_6 of 1. 4 mg/day, human milk was found to contain 0. 12 mg/l, and plasma PLP of nursing infants averaged 54 nmol. Extrapolation on the basis of metabolic body sise, weight, and growth suggests 0. 3 mg/day as an adequate intake for infants 6-12 months of age. Information on vitamin B_6 requirements for children is limited, but an age-related decrease in erythrocyte PLP and an increase in the aspartate aminotransferase activation.

In a review of earlier studies of men with various protein intakes, Linkswiler concluded that normalisation of a tryptophan load test required 1-1. 5 mg vitamin B_6. 1. 6 mg vitamin B_6 led to plasma PLP levels above 30 nmol/l for young men with various protein intakes. From several investigations of young women, a requirement closer to 1-1. 2 mg vitamin B_6 could be estimated.

Limited studies of the elderly indicate that requirements may be somewhat higher, at least to maintain plasma PLP above a 20-nmolar cut-off level.

For pregnancy it was confirmed that indicators of vitamin B_6 status decrease, especially in the third trimester. It is not clear, however, whether this is a normal physiologic phenomenon. For a maternal body store of 169 mg and foetal plus placental accumulation of 25 mg vitamin B_6, about 0. 1 mg/day is needed on average over gestation. With additional allowances for the increased metabolic need and weight of the mother and about 75 percent of bio-availability, an additional average requirement in pregnancy of 0. 25 mg can be assumed. Because most of this need is in the latter stages of pregnancy and vitamin B_6 is not stored to any significant extent, an extra 0. 5 mg/day of vitamin B_6 may be justified to err on the side of safety.

For lactation, it may be prudent to add 0. 6 mg vitamin B_6 to the base requirement for women because low maternal intakes could lead to a compromised vitamin B_6 status in the infant.

Table: Recommended nutrient intakes for vitamin B_6

Group	Recommended nutrient intake mg/day
Infants and children	
0-6 months	0. 1
7-12 months	0. 3
1-3 years	0. 5
4-6 years	0. 6
7-9 years	1. 0

Adolescents,	
10-18 years	1. 2 1. 3
Females	
Males	
Adults	
Females, 19-50 years	1. 3
Males, 19-50 years	1. 3
Females, >50 years	1. 5
Males, >50 years	1. 7
Pregnancy	1. 9
Lactation	2. 0

PANTOTHENATE

Deficiency

The widespread occurrence of releasable pantothenic acid in food makes a dietary deficiency unlikely. If a deficiency occurs, it is usually accompanied by deficits of other nutrients. The use of experimental animals, an antagonistic analogue (w-methyl-pantothenate) given to humans, and more recently the feeding of semi-synthetic diets virtually free of pantothenate have all helped to define signs and symptoms of deficiency. Subjects become irascible; develop postural hypotension, rapid heart rate on exertion, epigastric distress with anorexia and constipation, numbness and tingling of the hands and feet ("burning feet" syndrome); and have hyperactive deep tendon reflexes and weakness of finger extensor muscles. Some cases of pantothenate deficiency have been observed in patients with acne and other dermatitic conditions.

Functions

Pantothenic acid is a component of CoA, a cofactor that carries acyl groups for many enzymatic processes, and of phosphopantetheine within acyl carrier protein, a component of the fatty acid synthase complex. Pantothenate is most especially involved in fatty acid metabolism but has a wide-ranging function as a prosthetic group that adds specificity to binding with appropriate enzymes.

Biochemical Indicators

Indicators used to estimate pantothenate requirements are urinary excretion and blood levels. Excretion rate reflects intake. Whole blood, which contains vitamin and pantothenate-containing metabolites, has a general correlation with intake; erythrocyte levels seem more meaningful than plasma or serum levels.

Relative correspondence to pantothenate status has been reported for urinary excretion and for blood content of both whole blood and erythrocytes. A decline in urinary pantothenate levels from approximately 3-0. 8 mg/day (13. 7-3. 6 mmol/day) in young men fed a deficient diet for 84 days. Urinary excretion for a typical American diet was found to be 2. 6 mg (12mmol)/d, but it was strongly dependent on diet. Pantothenate intake estimated for adolescents was significantly correlated with pantothenate in urine.

Whole-blood pantothenate fell from 1. 95-1. 41 mg/ml (8. 8-6. 4 mmol/l)when six adult males were fed a pantothenate-free diet. Whole-blood content corresponded to intake, and the range in whole blood. 57-2. 66 mg/ml (7. 2- 12. 1mmol/l. There is an excellent correlation of whole-blood concentrations of pantothenate with the erythrocyte concentration, with an average value being 334 ng/ml (1. 5 mmol/l). The lack of sufficient population data, however, suggests the current use of an adequate intake rather than a recommended intake as a suitable basis for recommendations.

Factors Affecting Requirements

A measurement of urinary excretion of pantothenate after feeding a formula diet containing both bound and free vitamin indicates that approximately 50 percent of the pantothenate present in natural foods may be bio-available.

Findings by Age and Life Stage

Infant requirements are based on an estimation of pantothenic acid content of human milk, which according to reported values is approximately 2 mg/l. For a reported average human-milk intake of 0. 75 l/day these values suggest that 1. 6 mg/day is an adequate intake by the younger (0-6 months) infants. Taking into consideration growth and body sise, 1. 9 mg/day may be extrapolated for the older (6-12 months) infant.

A range of pantothenate intake of 4 to approximately 8 mg/day in 12 adolescents who were 11-16 years old. The usual pantothenate intake for American adults has been reported to be 4-7 mg/day. Hence, around 5 mg/day is apparently adequate.

For pregnancy there is only one relatively recent study that found lower blood pantothenate levels but no difference in urinary excretion in pregnant women compared with non-pregnant controls.

For lactation blood pantothenate concentrations were found significantly lower at 3 months post-partum. With a loss of 1. 7 mg/day (7. 8 mmol/d) from a lactating woman and lower maternal blood concentrations found with intakes of about 5-6 mg/d, a recommended intake may be 7 mg/d.

Table: Recommended nutrient intakes for pantothenate

Group	Recommended nutrient intake, mg/day
Infants and children	
0-6 months	1. 7
7-12 months	1. 8
1-3 years	2. 0
4-6 years	3. 0
7-9 years	4. 0
Adolescents,	
10-18 years	5. 0
Adults	
Females, 19+	5. 0
Males, 19+	5. 0
Pregnancy	6. 0
Lactation	7. 0

BIOTIN

Deficiency

Biotin deficiency in humans has been clearly documented with prolonged consumption of raw egg whites, which contain biotin-binding avidin. Biotin deficiency was also observed in cases of parenteral nutrition with solutions lacking biotin given to patients with short-gut syndrome and other causes of malabsorption. Some cases of biotin deficiency were noted in infants with intractable diaper dermatitis and in those fed special formulas. Dietary deficiency in otherwise normal people is probably rare. Some patients have multiple carboxylase deficiencies and there are occasional biotinidase deficiencies. Clinical signs of deficiency include dermatitis of an erythematous and seborrheic type; conjunctivitis; alopecia; and central nervous system abnormalities such as hypotonia, lethargy, and developmental delay in infants and depression, hallucinations, and paresthesia of the extremities in adults.

Toxicity

Toxicity is not a problem because of limited intestinal absorption of biotin.

Functions

Biotin functions as a co-enzyme within several carboxylases after the carboxyl function of the vitamin becomes amide linked to the e-amino of specific lysyl residues of the apoenzymes. In humans and other mammals, biotin operates within four carboxylases. Three of the four biotin-dependent carboxylases are mitochondrial (pyruvate carboxylase, methylcrotonyl-CoA carboxylase, and propionyl-CoA carboxylase) whereas the fourth (acetyl-CoA carboxylase) is found both in mitochondria and the cytosol. In all these cases biotin serves as carrier for the transfer of active bicarbonate into a substrate to generate a carboxyl product.

Biochemical Indicators

Indicators used to estimate biotin requirements are urinary excretion and 3-hydroxyisovalerate excretion. The excretion rate of vitamin and metabolites in urine is assessed by avidin-based radioimmunoassay with HPLC. 3-Hydroxyisovalerate excretion inversely reflects the activity of b-methyl-crotonyl-CoA carboxylase, which is involved in leucine metabolism.

The present indicators for biotin status are its urinary excretion, as assessed with an avidin-based radioimmunoassay with HPLC, and 3-hydroxyisovalerate excretion. The isolation and chemical identification of more than a dosen metabolites of biotin established the main features of its use in microbes and mammals. Quantification of the major biotin metabolites was done. Both biotin and bisnorbiotin excretions decline in parallel in individuals on a diet containing raw egg whites. In these individuals the levels of urinary 3-hydroxyisovalerate, which increase as a result of decreased activity of b-methylcrotonyl-CoA carboxylase and altered leucine metabolism, rose from a normal mean of 112 to 272 μmol/24 hours. Decreased excretion of biotin, abnormally increased excretion of 3-hydroxyisovalerate, or both have been reported for overt cases of biotin deficiency. The lack of sufficient population data, however, suggests the current use of an adequate intake rather than a recommended intake as a suitable basis for recommendations.

Findings by Age and Life Stage

The biotin content of human milk is estimated to be approximately 6 μg (24nmol)/l based on several studies that report values ranging from near 4-7 μg /l. Hence, the estimated intake of biotin for an infant consuming 0. 75 l is 5 μg/day during the first half year and for older infants is perhaps 6 μg/day.

Requirements for children and adults have been extrapolated as follows:

Adequate intake for child or adult = (adequate intake young infant) (weight adult or child/weight infant) $^{0.\ 75}$

For pregnancy there are at present insufficient data to justify an increase in the adequate intake, although decreased urinary biotin and 3-hydroxyisovalerate in a large fraction of seemingly healthy pregnant women. For lactation the intake may need to be increased by an additional 5 μg/day to cover the losses due to human-milk secretion.

Table: Recommended nutrient intakes for biotin

Group	Recommended nutrient intake mg/day
Infants and children	
0-6 months	5
7-12 months	6
1-3 years	8
4-6 years	12
7-9 years	20
Adolescents, 10-18 years	25
Adults	
Females, 19+	30
Males, 19+	30
Pregnancy	30
Lactation	35

GENERAL CONSIDERATIONS FOR B-COMPLEX VITAMINS

The recommendations for infants are based largely on the composition and quantity of human milk consumed and are formally considered to be adequate intakes. Younger (0-6 months) infants are considered to derive adequate intake from milk alone; recommendations for older (7-12 months) infants are adjusted by metabolic scaling such that a factor - (weight of 7-12 mo/weight of 0-5 mo) is multiplied by the recommendation for the younger infant. Recommendations have been given to use the higher level for the first year of life.

For most of the B vitamins, there is little or no direct information that can be used to estimate the amounts required by children and adolescents. Hence, an extrapolation from the adult level has been used where a factor - (weight child/

weight adult) (1 + growth factor) - is multiplied by the adult recommendation. For most of the B-complex vitamins covered here, data are not sufficient to justify altering recommendations for the elderly.

Dietary sources of B-complex vitamins

A listing of some better and usual food sources for the vitamins considered is given in Table.

Table: Dietary sources of water-soluble vitamins

Vitamin	Good-to-moderate dietary sources[a]
Thiamin (B_1)	Pork, organ meats, whole grains, and legumes
Riboflavin (B_2)	Milk and dairy products, meats, and green vegetables
Niacin (nicotinic acid and nicotinamide)	Liver, lean meats, grains, and legumes; can be formed from tryptophan
Vitamin B_6 (pyridoxine, pyridoxamine, and pyridoxal)	Meats, vegetables, and whole-grain cereals
Pantothenic acid	Animal tissues, whole-grain cereals, and legumes; widely distributed
Biotin	Liver, yeast, egg, yolk, soy flour, and cereals

Foods are listed according to the concentrations of vitamin which they contain.

Chapter 4

Role of Vitamins

VITAMIN A

ROLE OF VITAMIN A IN HUMAN

Vitamin A (retinol) is an essential nutrient needed in small amounts by humans for the normal functioning of the visual system; growth and development; and maintenance of epithelial cellular integrity, immune function, and reproduction. These dietary needs for vitamin A are normally provided for as preformed retinol (mainly as retinyl ester) and pro vitamin A carotenoids.

Overview of Vitamin A Metabolism

Preformed vitamin A in animal foods occurs as retinyl esters of fatty acids in association with membrane-bound cellular lipid and fat-containing storage cells. Pro-vitamin A carotenoids in foods of vegetable origin are also associated with cellular lipids but are embedded in complex cellular structures such as the cellulose-containing matrix of chloroplasts or the pigment-containing portion of chromoplasts. Normal digestive processes free vitamin A and carotenoids from embedding food matrices, a more efficient process from animal than from vegetable tissues. Retinyl esters are hydrolysed and the retinol and freed carotenoids are incorporated into lipid-containing, water-miscible micellar solutions. Products of fat digestion (e. g., fatty acids, monoglycerides, cholesterol, and phospholipids) and secretions in bile (e. g., bile salts and hydrolytic enzymes) are essential for the efficient solubilisation of retinol and especially for solubilisation of the very lipophilic carotenoids (e. g., aa- and bb-carotene, bb-cryptoxanthin, and lycopene) in the aqueous intestinal milieu.

Micellar solubilisation is a prerequisite to their efficient passage into the lipid-rich membrane of intestinal mucosal cells (i. e., enterocytes). Diets critically low in dietary fat (under about 5-10 g daily) or disease conditions that interfere with normal digestion and absorption leading to steatorrhea (e. g., pancreatic and liver diseases and frequent gastroenteritis) can therefore impede the efficient absorption of retinol and carotenoids. Retinol and some carotenoids enter the intestinal mucosal brush border by diffusion in accord with the concentration gradient between the micelle and plasma membrane of enterocytes.

Some carotenoids pass into the enterocyte and are solubilised into chylomicrons without further change whereas some of the pro-vitamin A carotenoids are converted to retinol by a cleavage enzyme in the brush border. Retinol is trapped intracellularly by re-esterification or binding to specific intracellular binding proteins. Retinyl esters and unconverted carotenoids together with other lipids are incorporated into chylomicrons, excreted into intestinal lymphatic channels, and delivered to the blood through the thoracic duct.

Tissues extract most lipids and some carotenoids from circulating chylomicrons, but most retinyl esters are stripped from the chylomicron remnant, hydrolysed, and taken up primarily by parenchymal liver cells. If not immediately needed, retinol is re-esterified and retained in the fat-storing cells of the liver (variously called adipocytes, stellate cells, or Ito cells). The liver parenchymal cells also take in substantial amounts of carotenoids. Whereas most of the body's vitamin A reserve remains in the liver, carotenoids are also deposited elsewhere in fatty tissues throughout the body. Usually, turnover of carotenoids in tissues is relatively slow, but in times of low dietary carotenoid intake, stored carotenoids are mobilised. A recent study in one subject using stable isotopes suggests that retinol can be derived not only from conversion of dietary pro-vitamin carotenoids in enterocytes - the major site of bioconversion - but also from hepatic conversion of circulating pro-vitamin carotenoids. The quantitative contribution to vitamin A requirements of carotenoid converted to retinoids beyond the enterocyte is unknown.

Following hydrolysis of stored retinyl esters, retinol combines with a plasma-specific transport protein, retinol-binding protein (RBP). This process, including synthesis of the unoccupied RBP (apo-RBP), occurs to the greatest extent within liver cells but it may also occur in some peripheral tissues. The RBP-retinol complex (holo-RBP) is secreted into the blood where it associates with another hepatically synthesised and excreted larger protein, transthyretin.

The transthyretin-RBP-retinol complex circulates in the blood, delivering the lipophilic retinol to tissues; its large sise prevents its loss through kidney filtration. Dietary restriction in energy, proteins, and some micronutrients can limit hepatic synthesis of proteins specific to mobilisation and transport of vitamin A. Altered kidney functions or fever associated with infections (e. g., respiratory infections [6] or diarrhoea [7]) can increase urinary vitamin A loss.

Holo-RBP transiently associates with target-tissue membranes, and specific intracellular binding proteins then extract the retinol. Some of the transiently sequestered retinol is released into the blood unchanged and is recycled. A limited reserve of intracellular retinyl esters is formed, that subsequently can provide functionally active retinol and its oxidation products (i. e., isomers of retinoic acid) as needed intracellularly.

These biologically active forms of vitamin A are associated with specific cellular proteins which bind with retinoids within cells during metabolism and with nuclear receptors that mediate retinoid action on the genome. Retinoids modulate the transcription of several hundreds of genes. In addition to the latter role of retinoic acid, retinol is the form required for functions in the visual and reproductive systems and during embryonic development.

Holo-RBP is filtered into the glomerulus but recovered from the kidney tubule and recycled. Normally vitamin A leaves the body in urine only as inactive metabolites which result from tissue utilisation and as potentially recyclable active glucuronide conjugates of retinol in bile secretions. No single urinary metabolite has been identified which accurately reflects tissue levels of vitamin A or its rate of utilisation. Hence, at this time urine is not a useful biologic fluid for assessment of vitamin A nutriture.

Biochemical Mechanisms for Vitamin A Functions

Vitamin A functions at two levels in the body. The first is in the visual cycle in the retina of the eye; the second is in all body tissues systemically to maintain growth and the soundness of cells. In the visual system, carrier-bound retinol is transported to ocular tissue and to the retina by intracellular binding and transport proteins. Rhodopsin, the visual pigment critical to dim-light vision, is formed in rod cells after conversion of all-trans retinol to retinaldehyde, isomerization to the 11-cis-form, and binding to opsin. Alteration of rhodopsin through a cascade of photochemical reactions results in ability to see objects in dim light. The speed at which rhodopsin is regenerated relates to the availability of retinol. Night blindness is usually an indicator of inadequate available retinol, but it can also be due to a deficit of other nutrients, which are critical to the regeneration of rhodopsin, such as protein and zinc, and to some inherited diseases, such as retinitis pigmentosa.

The growth and differentiation of epithelial cells throughout the body are especially affected by vitamin A deficiency (VAD). Goblet cell numbers are reduced in epithelial tissues. The consequence is that mucous secretions with their antimicrobial components diminish. Cells lining protective tissue surfaces fail to regenerate and differentiate, hence flatten and accumulate keratin. Both factors - the decline in mucous secretions and loss of cellular integrity - diminish resistance to invasion by potentially pathogenic organisms. The immune system is also compromised by direct interference with production of some types of protective secretions and cells. Classical symptoms of xerosis (drying or nonwetability) and desquamation of dead surface cells as seen in ocular tissue (i. e., xerophthalmia) are the external evidence of the changes also occurring to various degrees in internal epithelial tissues.

Current understanding of the mechanism of vitamin A action within cells outside the visual cycle is that cellular functions are mediated through specific nuclear receptors. These receptors are activated by binding with specific isomers of retinoic acid (i. e., all-trans and 9-cis retinoic acid). Activated receptors bind to DNA response elements located upstream of specific genes to regulate the level of expression of those genes. The synthesis of a large number of proteins vital to maintaining normal physiologic functions is regulated by these retinoid-activated genes. There also may be other mechanisms of action that are as yet undiscovered.

POPULATION AT RISK AND CONSEQUENCES OF VITAMIN A DEFICIENCY

VAD is not simply defined. WHO defines it as tissue concentrations of vitamin A low enough to have adverse health consequences even if there is no evidence of clinical xerophthalmia. In addition to the specific signs and symptoms of xerophthalmia and the risk of irreversible blindness, non-specific symptoms include increased morbidity and mortality, poor reproductive health, increased risk of anaemia, and contributions to slowed growth and development. Because these non-specific adverse consequences may occur from other nutrient deficits as well, it is difficult to attribute non-ocular symptoms specifically to VAD in the absence of biochemical measurements reflective of vitamin A status.

VAD can occur in individuals of any age. However, it is a disabling and potentially fatal public health problem for children under 6 years of age. VAD-related blindness is most prevalent in children under 3 years of age. This period is characterised by high requirements for vitamin A to support early rapid growth, the transition from breast-feeding to dependence on other dietary sources of the vitamin, and increased frequency of respiratory and gastrointestinal infections.

The increased mortality risk from concurrent infections extends at least to 6 years of age and is associated with both clinical and sub-clinical VAD. There is little information regarding the health consequences of VAD in school-age children. The prevalence of Bitot's spots (i. e., white foamy patches on the conjunctiva) may be highest in this age group but their occurrence may reflect past more than current history of VAD. Women of reproductive age are thought also to be vulnerable to VAD during pregnancy and lactation because they often report night blindness and because their breast-milk is frequently low in vitamin A. Not all night blindness in pregnant women, however, responds to vitamin A.

There is no consistent, clear indication in humans of a gender differential in vitamin A requirements during childhood. Growth rates and presumably the need for vitamin A from birth to 10 years for boys are consistently higher than those for girls. In the context of varied cultural and community settings, however, variations in gender-specific child-feeding and care practices are likely to subsume a small gender differential in requirements to account for reported gender differences and prevalence of xerophthalmia. Pregnant and lactating women require additional vitamin A to support maternal and foetal tissue growth and lactation losses, additional vitamin A which is not needed by other post-adolescent adults.

Risk factors

VAD is most common in populations consuming most of their vitamin A needs from pro-vitamin carotenoid sources and where minimal dietary fat is available. About 90 percent of ingested preformed vitamin A is absorbed, whereas the absorption efficiency of pro-vitamin A carotenoids varies widely depending on the type of plant source and the fat content of the accompanying meal. Where it is possible to increase dietary fat, this will likely improve the absorption of vitamin A activity from the diet.

In areas with endemic VAD, fluctuations in the incidence of VAD throughout the year reflect the balance between intake and need. Periods of general food shortage (and specific shortages in vitamin A-rich foods), peak incidence of common childhood infectious diseases (diarrhoea, respiratory infections, and measles), and periodic seasonal growth spurts affect the balance. Seasonal food availability can influence VAD prevalence by directly influencing access to pro-vitamin A sources; for example, the scarcity of mangoes in hot arid months followed by the glutting of the market with mangoes during harvest seasons. Seasonal growth spurts in children frequently follow seasonal postharvest increases in energy and macronutrient intakes. These increases are usually obtained from staple grains (e. g., rice) and tubers (e. g., light-coloured yams) which are not good sources of some micronutrients (e. g., vitamin A to support the growth spurt).

Food habits and taboos often restrict consumption of potentially good food sources of vitamin A (e. g., mangoes and green leafy vegetables). Culture-specific factors for feeding children, adolescents, and pregnant and lactating women are common. Illness- and childbirth-related proscription of the use of specific foods pervade in many traditional cultures. Such influences alter short- and long-term food distribution within families. However, some cultural practices can be protective of vitamin A status and they need to be identified and reinforced.

Morbidity and mortality

The consequences of VAD are manifested differently in different tissues. In the eye, the symptoms and signs, together referred to as xerophthalmia, have a long, well-recognised history and have until recently been the basis for estimating the global burden from the disease. Although ocular symptoms and signs are the most specific indicators of VAD, they occur only after other tissues have impaired functions that are less specific and less easily assessed.

The prevalence of ocular manifestations (i. e., xerophthalmia or clinical VAD) is now recognised to far underestimate the magnitude of the problem of functionally significant VAD. Many more preschool-age children and perhaps older children and women who are pregnant or lactating have their health compromised when they are sub-clinically deficient. In young children, sub-clinical deficiency, like clinical deficiency, increases the severity of some infections, particularly diarrhoea and measles, and the risk of dying.

The incidence and prevalence of diarrhoea may also increase with sub-clinical VAD. Meta-analyses conducted by three independent groups using data from several randomised trials provide convincing evidence that community-based improvement of the vitamin A status of deficient children 6 months to 6 years of age reduces their risk of dying by 20-30 percent on average. Mortality in children who are blind from keratomalacia or who have corneal disease is reported to be from 50 percent to 90 percent and measles mortality associated with VAD is increased up to 50 percent. Limited data are available from controlled studies of the possible link between morbidity history and vitamin A status of pregnant and lactating women.

There are discrepancies in the linkage between incidence and severity of infectious morbidity of various aetiologies and vitamin A status. The weight of evidence supports an association of VAD with severity of an infection once acquired, except for respiratory diseases, which are nonresponsive. The severity of the pneumonia associated with measles, however, is an exception because it decreases with treatment with vitamin A supplements.

Infectious diseases depress circulating retinol and contribute to vitamin A depletion. Enteric infections may alter absorptive-surface area, compete for absorption-binding sites, and increase urinary loss. Febrile systemic infections also increase urinary loss and metabolic utilisation rates and may reduce apparent retinol stores if fever occurs frequently. In the presence of latent deficiency, disease occurrence is often associated with precipitating ocular signs. Measles virus infection is especially devastating to vitamin A metabolism, adversely interfering with both efficiencies of utilisation and conservation. Severe protein-energy malnutrition affects many aspects of vitamin A metabolism, and even when some retinyl ester stores are still present, malnutrition - often coupled with infection - can prevent transport-protein synthesis, that results in immobilisation of existing vitamin A stores.

The compromised integrity of the epithelium, together with the possible alteration in hormonal balance at severe levels of deficiency, impairs normal reproductive functions in animals. Controlled human studies are, of course, lacking. In animals and Humans, congenital anomalies can result if the foetus is exposed to severe deficiency or large excesses of vitamin A at critical periods early in gestation (first trimester) when foetal organs are being formed. Reproductive performance measured by infant outcomes in one community-based clinical intervention trial, however, was not influenced by vitamin A status.

The growth of children may be impaired by VAD. Interventions with vitamin A only have not consistently demonstrated improved growth in community studies because VAD seldom occurs in isolation of other nutrient deficiencies that also affect growth and may be more limiting.

A lack of vitamin A can affect iron metabolism when deficiencies of both nutrients coexist and particularly in environments that favour frequent infections. Maximum haemoglobin response occurs when iron and vitamin A deficiencies are corrected together. VAD appears to influence the availability of storage iron for use by haematopoietic tissue.

However, additional research is needed to clarify the mechanisms of the apparent interaction.

UNITS OF EXPRESSION

In blood, tissues, and human milk, vitamin A levels are conventionally expressed in μg/dL or μmol/l of all-trans retinol. Except for postprandial conditions, most of the circulating vitamin A is retinol whereas in most tissues (such as the liver), secretions (such as human milk), and other animal food sources it exists mainly as retinyl esters, that are usually hydrolysed before analytical detection.

To express the vitamin A activity of carotenoids in diets on a common basis, a joint FAO/WHO Expert Group in 1967 introduced the concept of the retinol equivalent (RE) and established the following relationships among food sources of vitamin A:

1 μg retinol= 1 RE
1 μg b-carotene = 0. 167 μg RE
1 μg other pro-vitamin A carotenoids = 0. 084 μg RE

These equivalencies were derived from balance studies to account for the less-efficient absorption of carotenoids (thought to be about one-third that of retinol) and their bioconversion to vitamin A (one-half for b-carotene and one-fourth for other pro-vitamin carotenoids). It was recognised at the time that the recommended conversion factors (i. e., 1:6 for vitamin A:b-carotene and 1:12 for vitamin A: all other pro-vitamin carotenoids) were only average estimates for a mixed diet.

Recently there has been renewed interest in examining bio-availability factors by using more quantitative stable isotope techniques for measuring whole-body stores in response to controlled intakes and by following postabsorption carotenoids in the triacylglycerol-rich lipoprotein fraction. The data are inconsistent but in general suggests that revision toward lower bio-availability estimates is likely. Until additional definitive data are available, however, the above conversion factors will be used.

Retinol equivalents in a diet are calculated as the sum of the weight of the retinol portion of preformed vitamin A with the weight of b-carotene divided by its conversion factor and with the weight of other carotenoids divided by their conversion factor. Most recent food composition tables report b-carotene and sometimes other pro-vitamin A carotenoids as μg/g edible portion. However, older food composition tables frequently report vitamin A as international units (IUs). The following applies to determining comparable values as μg:

1 IU retinol = 0. 3 μg retinol
1 IU b-carotene = 0. 6 μg b-carotene
1 IU retinol = 3 IU b-carotene

It is strongly recommended that weight or molar units replace the use of IU to decrease confusion and overcome limitations in the non-equivalence of the IU values for retinol and beta-carotene.

DIETARY SOURCES

Preformed vitamin A is found almost exclusively in animal products, such as human milk, glandular meats, liver and fish liver oils (especially), egg yolk, and whole milk and dairy products. Preformed vitamin A is also used to fortify processed foods, that may include sugar, cereals, condiments, fats, and oils. Pro-vitamin A carotenoids are found in green leafy vegetables (e. g., spinach, amaranth, and young leaves from various sources), yellow vegetables (e. g., pumpkins, squash, and carrots), and yellow and orange noncitrus fruits (e. g., mangoes, apricots, and papaya). Red palm oil produced in several countries worldwide is especially rich in pro-vitamin A. Some other indigenous plants also may be unusually rich sources of pro-vitamin A. Such examples are the palm fruit known in Brazil as buriti, that is found in areas along the Amazon, and the fruit known as gac in Vietnam, that is used to colour rice, particularly on ceremonial occasions. Foods containing pro-vitamin A carotenoids tend to be less biologically available but more affordable than animal products. It is mainly for this reason that carotenoids provide most of the vitamin A activity in the diets of economically deprived populations.

Dietary Intake and Patterns

Vitamin A status cannot be assessed from dietary intake alone, but dietary intake assessment can provide evidence of risk of an inadequate status. Quantitative collection of dietary information is fraught with measurement problems. These problems arise both from obtaining representative quantitative dietary histories from individuals, communities, or both and from interpreting these data while accounting for differences in bio-availability, preparation losses, and variations in food composition data among population groups. This is especially difficult in populations consuming most of their dietary vitamin A from pro-vitamin carotenoid sources. Simplified guidelines have been developed recently in an effort to improve the obtaining of reliable dietary intake information from individuals and communities.

World and Regional Supply and Patterns

In theory the world's food supply is sufficient to meet global requirements. Great differences exist, however, in the available sources (animal and vegetable) and in per capita consumption of the vitamin among different countries, age categories, and socio-economic groups. VAD as a global public health problem, therefore, is largely due to inequitable food distribution among and within countries and households in relation to need for ample bio-available vitamin A sources.

Earlier FAO global estimates in 1984 indicated that preformed vitamin A constituted about one-third of total dietary vitamin A. World availability of vitamin A for human consumption at that time was approximately 220 μg of preformed retinol per capita daily and 560 μg RE from pro-vitamin carotenoids per person per day, for a total of about 790 μg RE. These values are based on supply estimates and not consumption estimates. Losses commonly occur during food storage and processing, both industrially and in the home.

The estimated available regional supply of vitamin A from a more recent global evaluation the variability in amounts and sources of vitamin A. The variability is further complicated by access to the available supply, that varies with household income, poverty being a yardstick for risk of VAD. VAD is most prevalent in Southeast Asia, Africa, and the Western Pacific, where vegetable sources contribute nearly 80 percent or more of the available supply of retinol equivalents. Furthermore, in Southeast Asia the total available supply is about half of that of most other regions and is particularly low in animal sources.

EVIDENCE FOR MAKING RECOMMENDATIONS

Indicators of Vitamin A Deficiency

Ocular signs of VAD are measured by clinical examination and history and are quite specific in preschool-age children. However, these are rare occurrences that require examination of large populations to obtain incidence and prevalence data. Sub-clinical VAD is more prevalent, requiring smaller sample sises to obtain valid prevalence estimates.

A full description of clinical indicators with coloured illustrations for each is found in the WHO Field Guide. The most frequently occurring is night blindness, which is the earliest manifestation of xerophthalmia. In mild form it is generally noticeable after stress from a bright light that bleaches the rhodopsin (visual purple) found in the retina. VAD prolongs the time to regenerate rhodopsin, thus delays adaptation time in dark environments. Night-blind young children tend to stumble when going from bright to dimly lighted areas and they, as well as night-blind mothers, tend to remain inactive at dusk and at night.

No field-applicable objective tool is currently available for measuring night blindness in children under about 3 years of age. It can be measured by history in certain cultures. When night blindness is prevalent, many cultures coin a word descriptive of the characteristic symptom that they can reliably recall on questioning, making this a useful tool for assessing the prevalence of VAD.

Questioning for night blindness is not consistently a reliable assessment measure where a local term is absent. In addition, there is no clearly defined blood retinol level that is directly associated with occurrence of the symptom. Vitamin A - related night blindness, however, responds rapidly, usually within 1-2 days, to administration of vitamin A.

Sub-clinical Vitamin A deficiency

Direct measurement of concentrations of vitamin A in the liver where it is stored or in the total body pool relative to known specific vitamin A-related functions (e. g., night blindness) would be the indicator of choice for determining requirements. This cannot be done with the methodology now available for population use. There are several practical biochemical methods for estimating sub-clinical vitamin A status but all have limitations. Each method is useful to identify deficient populations, but no one of these indicators is definitive or directly related quantitatively to disease occurrence. The indicators of choice are listed in Table.

These indicators are less specific to VAD than clinical eye signs and less sensitive for measuring sub-clinical vitamin A status. WHO recommends that where feasible at least two sub-clinical biochemical indicators, or one biochemical and a composite of non-biochemical risk factors, should be measured and that both types of indicators should point to deficiency in order to identify populations at high risk of VAD. Cut-off points given in Table represent the consensus gained from practical experience comparing populations with some evidence of VAD with those without VAD.

There are no field studies that quantitatively relate the prevalence of adverse health symptoms (e. g., incidence or prevalence of severe diarrheal disease) and relative levels of biologic indicator cut-off values. Furthermore, each of the biochemical indicators listed is subject to confounding factors, which may be unrelated to vitamin A status (e. g., infections).

Table Biochemical indicators of sub-clinical VAD in mothers and in children 6-71 months of age

Although all biochemical indicators currently available have limitations, the biochemical indicator of choice for population assessment is the distribution of serum levels of vitamin A (serum retinol). Only at very low blood levels (<0. 35 mmol/l) is there an association with corneal disease prevalence. Blood levels between 0. 35 and 0. 70 mmol/l are likely to characterise sub-clinical deficiency, but sub-clinical deficiency may still be present at levels between 0. 70 and 1. 05 mmol/l and occasionally above 1. 05 mmol/l.

The prevalence of values below 0. 70 mmol/l is a generally accepted population cut-off for preschool-age children to indicate risk of inadequate vitamin A status and above 1. 05 mmol/l to indicate an adequate status. As noted elsewhere, clinical and sub-clinical infections can lower serum levels of vitamin A on average as much as 25 percent independently of vitamin A intake. Therefore, at levels between about 0. 5 and 1. 05 mmol/l, the relative dose response or modified relative dose response test on a subsample of the population can be useful for identifying the prevalence of critically depleted body stores when interpreting the left portion of serum retinol distribution curves.

DETERMINING MEAN REQUIREMENTS AND SETTING SAFE LEVELS OF INTAKE

Requirement and safe level of intake for vitamin A in this report do not differ significantly from those of the 1988 FAO/WHO Expert Consultation except for adapting to the age categories defined in this consultation and during pregnancy. The term safe level of intake used in the 1988 report is retained in this report because the levels in Table do not strictly correspond to the definition of a recommended nutrient intake.

The mean requirement for an individual is defined as the minimum daily intake of vitamin A as presented in mmg retinol equivalents (mg RE) to prevent xerophthalmia in the absence of clinical or sub-clinical infection.

This intake should account for proportionate bio-availability of preformed vitamin (about 90 percent) and pro-vitamin A carotenoids from a diet that contains sufficient fat (e. g., at least 5-10 g). Bio-availability of carotenoids varies widely by source (e. g., fibrous green leafy vegetables or soft-tissue fruits). The required level of intake is set to prevent clinical signs of deficiency, allow for normal growth, and reduce the risk of vitamin A - related severe morbidity and mortality on a population basis. It does not allow for frequent or prolonged periods of infections or other stresses.

The safe level of intake for an individual is defined as the average continuing intake of vitamin A required to permit adequate growth and other vitamin A-dependent functions and to maintain an acceptable total body reserve of the vitamin. This reserve helps offset periods of low intake or increased need resulting from infections and other stresses.

Useful indicators include a plasma retinol concentration above 0. 70 mmol/l, that is associated with a relative dose response below 20 percent, or a modified relative dose response below 0. 06. For lactating women, breast-milk retinol levels above 1. 05 mmol/l (or above 8 mg/g milk fat) are considered to reflect minimal maternal stores because levels above 1. 75 mmol/l are common to populations known to be healthy and without evidence of insufficient dietary vitamin A.

Infants and Children

Vitamin A requirements for infants are calculated from the vitamin A provided in human milk. During at least the first 6 months of life, exclusive breast-feeding can provide sufficient vitamin A to maintain health, permit normal growth, and maintain sufficient stores in the liver.

Reported retinol concentrations in human milk varies widely from country to country (0. 70-2. 45 mmol/l). In some developing countries the vitamin A intake of breast-fed infants who grow well and do not show signs of deficiency is from 120 to 170 mg RE/day. Such intakes are considered adequate to cover infant requirements if the infant's weight is assumed to be at least at the 10th percentile according to WHO standards. However, this intake is unlikely to build adequate body stores because xerophthalmia is common in preschool-age children in the same communities with somewhat lower intakes. Because of the need for vitamin A to support the growth rate of infancy, which can vary considerably, a requirement estimate of 180 mg RE/day seems appropriate.

The safe level for infants up to 6 months of age is based on observations of breast-fed infants in communities in which good nutrition is the norm. Average consumption of human milk by such infants is about 750 ml/day during the first 6 months. Assuming an average concentration of vitamin A in human milk of about 1. 75 mmol/l, the mean daily intake would be about 375 mg RE, which is therefore the recommended safe level. From 7-12 months, human milk intake averages 650 ml, which would provide 325 μg vitamin A daily. Because breast-fed infants in endemic vitamin A-deficient populations are at increased risk of death from 6 months onward, the requirement and recommended safe intake are increased to 190 mg and 400 mg, respectively.

The requirement (with allowance for variability) and the recommended intake for older children may be estimated from those derived for late infancy. On this basis, including allowances for storage requirements and variability, requirements for preschool children would be in the range of 200-400 mg RE daily. In poor communities where children 1-6 years old are reported to have intakes of about 100-200 mg RE/day, signs of VAD do occur; in southern India these signs were relieved and risk of mortality was reduced when the equivalent of 350-400 mg RE/day was given to children weekly. In the United States most preschool-age children maintain serum retinol levels of 0. 70 mmol/l or higher while consuming diets providing 300-400 mg RE/day.

Adults

Estimates for the requirements and recommended safe intakes for adults are also estimated from those derived for late infancy, i. e. 4. 8 and 9. 3 mg RE/kg body weight/day. Detailed accounts for arriving at the requirement for vitamin A. The safe intakes recommended are consistent with the per capita vitamin A content in the food supply of countries that show adequate vitamin A status in all sectors of the population. Additional evidence that the existing safe level of intake is adequate for adults on a population basis is provided by an analysis of dietary data from the 1990 survey of British adults in whom there was no evidence of VAD. The median intake for another survey of nonpregnant UK women who did not consume liver or liver products during the survey week was 686 mg/day. This value is substantially above the estimated mean requirement for pregnant women and falls quite short of the amount in which teratology risk is reported. About one-third of the calculated retinol equivalents consumed by the British women came from pro-vitamin A sources.

Pregnancy

During pregnancy additional vitamin A is needed for the growth and maintenance of the foetus for providing a limited reserve in the foetal liver and for maternal tissue growth. There are no reliable figures available for the specific vitamin A requirements for these processes.

Newborn infants need around 100 mg retinol daily to meet their needs for growth. During the third trimester the foetus grows rapidly and, although obviously smaller in sise than the infant born full term, the foetus presumably has similar needs. Incremental maternal needs associated with pregnancy are assumed to be provided from maternal reserves in populations of adequately nourished healthy mothers. In populations consuming at the basal requirement, an increment of 100 mg/day during the full gestation period should enhance maternal storage during early pregnancy and allow adequate amounts of vitamin A for the rapidly growing foetus in late pregnancy. However, this increment may be minimal for women who normally ingest only the basal requirement level of vitamin A inasmuch as the needs and growth rate of the foetus will not be affected by the mother's initial vitamin A reserves.

Lactation

If the amounts of vitamin A recommended for infants are supplied by human milk, mothers should absorb at least as much in their diets to replace maternal losses. Thus, the increments in basal and recommended intakes during lactation are 180 mg RE and 350 mg RE, respectively, for the safe recommended intake per day. After the infant reaches the age of 6 months or when solid foods are introduced, the mother's need for additional amounts of vitamin A lessens.

Elderly

There is no indication that the vitamin A requirements of healthy elderly individuals differs from those of other adults. It should be remembered, however, that diseases that impede vitamin A absorption, storage, and transport might be more common in the elderly than in other age groups.

Table Estimated mean requirement and safe level of intake for vitamin A

Age group	Mean requirement µg RE/day	Recommended safe intake µg RE/day
Infants and children		
0-6 months	180	375
7-12 months	190	400
1-3 years	200	400
4-6 years	200	450
7-years	250	500
Adolescents, 10-18 years	330-400	600
Adults		
Females,19-65 years	270	500
Males, 19-65 years	300	600
65+	300	600
Pregnant women	370	800
Lactating women	450	850

TOXICITY

Because vitamin A is fat soluble and can be stored, primarily in the liver, routine consumption of large amounts of vitamin A over a period of time can result in toxic symptoms, including liver damage, bone abnormalities and joint pain, alopecia, headaches and vomiting, and skin desquamation. Hypervitaminosis A appears to be due to abnormal transport and distribution of vitamin A and retinoids caused by overloading of the plasma transport mechanisms.

The smallest daily supplement associated with liver cirrhosis that has been reported is 7500 μg taken for 6 years. Very high single doses can also cause transient acute toxic symptoms that may include bulging fontanels in infants; headaches in older children and adults; and vomiting, diarrhoea, loss of appetite, and irritability in all age groups. Rarely does toxicity occur from ingestion of food sources of preformed vitamin A. When this occurs, it usually results from very frequent consumption of liver products. Toxicity from food sources of pro-vitamin A carotenoids is not reported except for the cosmetic yellowing of skin.

Infants, including newborns, administered single doses equivalent to 15 000-30 000 mg retinol (50 000-100 000 IU) in oil generally show no adverse symptoms. However, daily prophylactic or therapeutic doses should not exceed 900 mg, that is well above the mean requirement of about 200 mg daily for infants. An excess of bulging fontanels occurred in infants under 6 months of age in one endemically deficient population given two or more doses of 7500 mg or 15 000 mg preformed vitamin A in oil, but other large-scale controlled clinical trials have not reported excess bulging after three doses of 7500 mg given with diptheria-pertussis-tetanus immunisations at about 6, 10, and 14 weeks of age. No effects were detected at 3 years of age that related to transient vitamin A-induced bulging that had occurred before 6 months of age.

Most children 1-6 years of age tolerate single oral doses of 60 000 mg (200 000 IU) vitamin A in oil at intervals of 4-6 months without adverse symptoms. Occasionally diarrhoea or vomiting is reported but is transient with no lasting sequelae. Older children seldom experience toxic symptoms unless they habitually ingest vitamin A in excess of 7500 mg (25 000 IU) for prolonged periods of time.

When taken by women at early stages of gestation at daily levels of more than 7500 mg (25 000 IU), foetal anomalies and poor reproductive outcomes are reported. One report suggests an increased risk of teratogenicity at intakes as low as 3000 mg (10 000 IU) but this is not confirmed by other studies. Women who are pregnant or might become pregnant should avoid taking excessive amounts of vitamin A. A careful review of the latest available information by a WHO Expert Group recommended that daily intakes in excess of 3000 mg (10 000 IU) or weekly intakes in excess of 7500 mg (25 000 IU), should not be taken at any period during gestation. High doses of vitamin A (60 000 mg, or 200 000 IU) can be safely given to breast-feeding mothers for up to 2 months postpartum and for 6 weeks for women who are not breast-feeding.

VITAMIN B_{12}

ROLE OF VITAMIN B_{12} IN HUMAN METABOLIC PROCESSES

Although the nutritional literature still uses the term vitamin B_{12}, a more specific name for vitamin B_{12} is cobalamin. Vitamin B_{12} is the largest of the B complex vitamins, with a molecular weight of over 1000. It consists of a corrin ring made up of four pyrroles with cobalt at the centre of the ring.

There are several vitamin B_{12}-dependent enzymes in bacteria and algae, but no species of plants have the enzymes necessary for vitamin B_{12} synthesis. This fact has significant implications for the dietary sources and availability of vitamin B_{12}. In mammalian cells there are only two vitamin B_{12}-dependent enzymes. One of these enzymes, methionine synthase, uses the chemical form of the vitamin which has a methyl group attached to the cobalt and is called methylcobalamin. The other enzyme, methylmalonyl CoA mutase, uses vitamin B_{12} with a 5′-adeoxyadenosyl moiety attached to the cobalt and is called 5′-deoxyaldenosylcobalamin, or coenzyme B_{12}.

In nature there are two other forms of vitamin B_{12}: hydroxycobalamin and aquacobalamin, where hydroxyl and water groups, respectively, are attached to the cobalt. The synthetic form of vitamin B_{12} found in supplements and fortified foods is cyanocobalamin, which has cyanide attached to the cobalt. These three forms of B_{12} are enzymatically activated to the methyl- or deoxyadeno-sylcobalamins in all mammalian cells.

DIETARY SOURCES AND AVAILABILITY

Most microorganisms, including bacteria and algae, synthesise vitamin B_{12}, and they constitute the only source of the vitamin. The vitamin B_{12} synthesised in microorganisms enters the human food chain through incorporation into food of animal origin. In many animals gastrointestinal fermentation supports the growth of these vitamin B_{12}-synthesising microorganisms, and subsequently the vitamin is absorbed and incorporated into the animal tissues. This is particularly true for the liver, where vitamin B_{12} is stored in large concentrations..

Products from these herbivorous animals, such as milk, meat, and eggs, constitute important dietary sources of the vitamin unless the animal is subsisting in one of the many regions known to be geochemically deficient in cobalt. Milk from cows and humans contains binders with very high affinity for vitamin B_{12}, whether they hinder or promote intestinal absorption is not entirely clear. Omnivores and carnivores, including humans, derive dietary vitamin B_{12} from animal tissues or products (i. e., milk, butter, cheese, eggs, meat, poultry, etc.). It appears that no significant amount of the required vitamin B_{12} by humans is derived from microflora, although vegetable fermentation preparations have also been reported as being possible sources of vitamin B_{12}.

ABSORPTION

The absorption of vitamin B_{12} in humans is complex. Vitamin B_{12} in food is bound to proteins and is released from the proteins by the action of a high concentration of hydrochloric acid present in the stomach. This process results in the free form of the vitamin, which is immediately bound to a mixture of glycoproteins secreted by the stomach and salivary glands. These glycoproteins, called R-binders (or haptocorrins), protect vitamin B_{12} from chemical denaturation in the stomach. The stomach's parietal cells, which secrete hydrochloric acid, also secrete a glycoprotein called intrinsic factor. Intrinsic factor binds vitamin B_{12} and ultimately enables its active absorption.

Although the formation of the vitamin B_{12} - intrinsic factor complex was initially thought to happen in the stomach, it is now clear that this is not the case. At an acidic pH the affinity of the intrinsic factor for vitamin B_{12} is low whereas its affinity for the R-binders is high. When the contents of the stomach enter the duodenum, the R-binders become partly digested by the pancreatic proteases, which causes them to release their vitamin B_{12}. Because the pH in the duodenum is more neutral than that in the stomach, the intrinsic factor has a high binding affinity to vitamin B_{12}, and it quickly binds the vitamin as it is released from the R-binders. The vitamin B_{12}-intrinsic factor complex then proceeds to the lower end of the small intestine, where it is absorbed by phagocytosis by specific ileal receptors.

POPULATIONS AT RISK FOR AND CONSEQUENCES OF VITAMIN B_{12} DEFICIENCY

Vegetarians

Because plants do not synthesise vitamin B_{12}, individuals who consume diets completely free of animal products (vegan diets) are at risk of vitamin B_{12} deficiency. This is not true of lacto-ovo-vegetarians, who consume the vitamin in eggs, milk, and other dairy products.

Pernicious Anaemia

Malabsorption of vitamin B_{12} can occur at several points during digestion. By far the most important condition resulting in vitamin B_{12} malabsorption is the auto-immune disease called pernicious anaemia (PA). In most cases of PA, antibodies are produced against the parietal cells causing them to atrophy, lose their ability to produce intrinsic factor, and secrete hydrochloric acid.

In some forms of PA the parietal cells remain intact but auto-antiobodies are produced against the intrinsic factor itself and attach to it, thus preventing it from binding vitamin B_{12}. In another less common form of PA, the antibodies allow vitamin B_{12} to bind to the intrinsic factor but prevent the absorption of the intrinsic factor-vitamin B_{12} complex by the ileal receptors. As is the case with most auto-immune diseases, the incidence of PA increases markedly with age. In most ethnic groups it is virtually unknown to occur before the age of 50, with a progressive rise in incidence thereafter.

However, African American populations are known to have an earlier age of presentation. In addition to causing malabsorption of dietary vitamin B_{12}, PA also results in an inability to reabsorb the vitamin B_{12} which is secreted in the bile. Biliary secretion of vitamin B_{12} is estimated to be between 0. 3 and 0. 5 μg/day.

Interruption of this so-called enterohepatic circulation of vitamin B_{12} causes the body to go into a significant negative balance for the vitamin. Although the body typically has sufficient vitamin B_{12} stores to last 3-5 years, once PA has been established the lack of absorption of new vitamin B_{12} is compounded by the loss of the vitamin because of negative balance. When the stores have been depleted, the final stages of deficiency are often quite rapid, resulting in death in a period of months if left untreated.

Atrophic Gastritis

Historically, PA was considered to be the major cause of vitamin B_{12} deficiency, but it was a fairly rare condition, perhaps affecting 1 percent to a few percent of elderly populations. More recently it has been suggested that a far more common problem is that of hypochlorhydria associated with atrophic gastritis, where there is a progressive reduction with age of the ability of the parietal cells to secrete hydrochloric acid. It is claimed that perhaps up to one-quarter of elderly subjects could have various degrees of hypochlorhydria as a result of atrophic gastritis.

It has also been suggested that bacterial overgrowth in the stomach and intestine in individuals suffering from atrophic gastritis may also reduce vitamin B_{12} absorption. This absence of acid is postulated to prevent the release of protein-bound vitamin B_{12} contained in food but not to interfere with the absorption of the free vitamin B_{12} found in fortified foods or supplements. Atrophic gastritis does not prevent the reabsorption of bilary vitamin B_{12} and therefore does not result in the negative balance seen in individuals with PA. However, it is agreed that with time, a reduction in the amount of vitamin B_{12} absorbed from the diet will eventually deplete even the usually adequate vitamin B_{12} stores, resulting in overt deficiency.

When considering recommended nutrient intakes (RNIs) for vitamin B_{12} for the elderly, it is important to take into account the absorption of vitamin B_{12} from sources such as fortified foods or supplements as compared with dietary vitamin B_{12}. In the latter instances, it is clear that absorption of intakes of less than 1. 5-2. 0 μg/day is complete - that is, for intakes of less than 1. 5-2. 0 μg of free vitamin B_{12}, the intrinsic factor - mediated system absorbs all of that amount. It is probable that this is also true of vitamin B_{12} in fortified foods, although this has not specifically been examined. However, absorption of food-bound vitamin B_{12} has been reported to vary from 9 percent to 60 percent depending on the study and the source of the vitamin, which is perhaps related to its incomplete release from food. This has led many to estimate absorption as being up to 50 percent to correct for bio-availability of absorption from food.

VITAMIN B_{12} INTERACTION WITH FOLATE OR FOLIC ACID

One of the vitamin B_{12} - dependent enzymes, methionine synthase, functions in one of the two folate cycles - the methylation cycle. This cycle is necessary to maintain availability of the methyl donor S-adenosylmethionine; interruption reduces the wide range of methylated products. One such important methylation is that of myelin basic protein. Reductions in the level of S-adenosylmethionine seen in PA and other causes of vitamin B_{12} deficiency produce demyelination of the peripheral nerves and the spinal column, called sub-acute combined degeneration. This neuropathy is one of the main presenting conditions in PA. The other principal presenting condition in PA is a megaloblastic anaemia morphologically identical to that seen in folate deficiency. Disruption of the methylation cycle should cause a lack of DNA biosynthesis and anaemia.

The methyl trap hypothesis is based on the fact that once the cofactor 5,10-methylenetetrahydrofolate is reduced by its reductase to form 5-methyltetrahydrofolate, the reverse reaction cannot occur. This suggests that the only way for the methyltetrahydrofolate to be recycled to tetrahydrofolate, and thus to participate in DNA biosynthesis and cell division, is through the vitamin B_{12} - dependent enzyme methionine synthase. When the activity of this synthase is compromised, as it would be in PA, the cellular folate will become progressively trapped as 5-methyltetrahydrofolate.

This will result in a cellular pseudo folate deficiency where despite adequate amounts of folate an anaemia will develop that is identical to that seen in true folate deficiency. Clinical symptoms of PA, therefore, include neuropathy, anaemia, or both. Treatment with vitamin B_{12}, if given intramuscularly, will reactivate methionine synthase, allowing myelination to restart. The trapped folate will be released and DNA synthesis and generation of red cells will cure the anaemia. Treatment with high concentrations of folic acid will treat the anaemia but not the neuropathy of PA.

It should be stressed that the so-called masking of the anaemia of PA is generally agreed not to occur at concentrations of folate found in food or at intakes of the synthetic form of folic acid found at usual RNI levels of 200 or 400 μg/day. However, there is some evidence that amounts less than 400 μg may cause a haematologic response and thus potentially treat the anaemia. The masking of the anaemia definitely occurs at high concentrations of folic acid (>1000 μg/day). This becomes a concern when considering fortification with synthetic folic acid of a dietary staple such as flour.

In humans the vitamin B_{12} - dependent enzyme methylmalonyl coenzyme A (CoA) mutase functions in the metabolism of propionate and certain of the amino acids, converting them into succinyl CoA, and in their subsequent metabolism via the citric acid cycle. It is clear that in vitamin B_{12} deficiency the activity of the mutase is compromised, resulting in high plasma or urine concentrations of methylmalonic acid (MMA), a degradation product of methylmalonyl CoA. In adults this mutase does not appear to have any vital function, but it clearly has an important role during embryonic life and in early development. Children deficient in this enzyme, through rare genetic mutations, suffer from mental retardation and other developmental defects.

ASSESSMENT OF VITAMIN B_{12} STATUS

Traditionally it was thought that low vitamin B_{12} status was accompanied by a low serum or plasma vitamin B_{12} level. Recently this has been challenged, who suggested that a proportion of people with normal vitamin B_{12} levels are in fact vitamin B_{12} deficient. They also suggested that elevation of plasma homo-cysteine and plasma MMA are more sensitive indicators of vitamin B_{12} status. Although plasma homo-cysteine may also be elevated because of folate or vitamin B_6 deficiency, elevation of MMA apparently always occurs with poor vitamin B_{12} status. There may be other reasons why MMA is elevated, such as renal insufficiency, so the elevation of itself is not diagnostic. Many would feel that low or decreased plasma vitamin B_{12} levels should be the first indication of poor status and that this could be confirmed by an elevated MMA if this assay was available.

EVIDENCE ON WHICH TO BASE A RECOMMENDED INTAKE

The Food and Nutrition Board of the National Academy of Sciences (NAS) Institute of Medicine has recently exhaustively reviewed the evidence of intake, status, and health for all age groups and during pregnancy and lactation. This review has lead to calculations of what they have called an estimated average requirement (EAR). The EAR is defined by NAS as "the daily intake value that is estimated to meet the requirement, as defined by the specific indicator of adequacy, in half of the individuals in a life-stage or gender group". They have estimated the recommended dietary allowances to be this figure plus 2 standard deviations (SDs). Some members of the Food and Agriculture Organization of the United Nations and World Health Organization (FAO/ WHO) Expert Consultation were involved in the preparation and review of the NAS recommendations and judge them to have been the best estimates based on available scientific literature. The FAO/WHO expert group felt it appropriate to adopt the same approach used by the NAS in deriving the RNIs. Therefore the RNIs suggested in Table are based on the NAS EARs plus 2 SDs.

Adults

Several lines of evidence point to an adult average requirement of about 2. 0 μg/day. The amount of intramuscular vitamin B_{12} needed to maintain remission in people with PA suggests a requirement of about 1. 5 μg/day, but they would also be losing 0. 3-0. 5μg/day through interruption of their enterohepatic circulation, which is not typical. This might suggest a requirement of 0. 7-1. 0 μg/day. Because vitamin B_{12} is not completely absorbed from food, an adjustment of 50 percent has to be added giving a range of 1. 4-2. 0 μg/day. Therapeutic response to ingested food vitamin B_{12} suggests a minimum requirement of something less than

1. 0 μg/day. Diets containing 1. 8 μg/day seemed to maintain adequate status but lower intakes showed some signs of deficiency. Dietary intakes of less than 1. 5 μg/day were reported to be inadequate in some subjects.

In summary, the average requirement could be said to be 2 μg/day. The variability of the requirements for vitamin B_{12} is accounted for by adding two SDs, that is, 2. 4 μg/day as an RNI for adults, including the elderly.

Table: Estimated average requirement (EAR) and recommended nutrient intake (RNI) for vitamin B_{12}, by age group

Group	EAR (μg/day)	RNI (μg/day)
Infants and children		
0-6 months	0. 32	0. 4
7-12 months	0. 32	0. 5
1-3 years	0. 7	0. 9
4-6 years	1. 0	1. 2
7-9 years	1. 5	1. 8
Adolescents, 10-18 years	2. 0	2. 4
Adults		
19-65 years	2. 0	2. 4
65+ years	2. 0	2. 4
Pregnancy	2. 2	2. 6
Lactation	2. 4	2. 8

Children

The Food and Nutrition Board of the NAS Institute of Medicine suggested the same intakes for adolescents with progressive reduction of intake for younger groups.

Pregnancy

Expert Consultation suggested that 0. 1-0. 2 μg/day of vitamin B_{12} is transferred to the foetus during the last two trimesters of pregnancy. On the basis of on foetal liver content from post-mortem samples, there is further evidence that the foetus accumulates on average 0. 1-0. 2 μg/day during pregnancies of women with diets which have adequate vitamin B_{12}. It has been reported that children born to vegetarians or other women with a low vitamin B_{12} intake subsequently develop signs of clinical vitamin B_{12} deficiency such as neuropathy. Therefore, when calculating the EAR for pregnant women, 0. 2 μg/day of vitamin B_{12} is added to the EAR for adults to result in an EAR of 2. 2 μg/day and an RNI of 2. 6 μg /day during pregnancy.

Lactation

It is estimated that 0. 4 μg/day of vitamin B_{12} is found in the human milk of women with adequate vitamin B_{12} status. Therefore an extra 0. 4 μg/day of vitamin B_{12} is needed during lactation in addition to the normal adult requirement of 2. 0 μg/day, giving a total EAR of 2. 4 μg/day during lactation and an RNI of 2. 8 μg/day.

Infants

As with other nutrients, the principal way to determine requirements of infants is to examine the levels in milk from mothers on adequate diets. There is a wide difference in the vitamin B_{12} values reported in human milk because of differences in methodology. The previous FAO/WHO report used milk vitamin B_{12} values of normal women of about 0. 4 μg/l. For an average milk production of 0. 75 l/day, the vitamin B_{12} intake by infants would be 0. 3 μg/day. Other studies have reported ranges of vitamin B_{12} in human milk to be 0. 4-0. 8 μg/L. Although daily intakes ranging from 0. 02 to 0. 05 μg/day have been found to prevent deficiency, these intakes are totally inadequate for long-term health. An EAR of 0. 3-0. 6 μg/day would result in an RNI of 0. 36-0. 72 μg/d. It might be prudent to use the lower figure of 0. 4 μg/day for the first 6 months of pregnancy and 0. 7 μg/day for last trimester.

Upper Limits

The absorption of vitamin B_{12} mediated by intrinsic factor is limited to 1. 5-2. 0 μg per meal because of the limited capacity of the receptors. In addition, between 1 percent and 3 percent of any particular oral administration of vitamin B_{12} is absorbed by passive diffusion. Thus, if 1000 μg vitamin B_{12} (sometimes used to treat those with PA) is taken orally, the amount absorbed would be 2. 0 μg by active absorption plus about 30 μg by passive diffusion. This amount has never been reported to have any side effects. Similar large amounts have been used in some preparations of nutritional supplements without apparent ill effects. However, there are no established benefits for such amounts. Such high intakes thus represent no benefit in those without malabsorption and should probably be avoided.

VITAMIN C

Vitamin C (chemical names: ascorbic acid and ascorbate) is a six-carbon lactone which is synthesised from glucose by many animals. Vitamin C is synthesised in the liver in some mammals and in the kidney in birds and reptiles. However, several species - including humans, non-human primates, guinea pigs, Indian fruit bats, and Nepalese red-vented bulbuls - are unable to synthesise vitamin C.

When there is insufficient vitamin C in the diet, humans suffer from the potentially lethal deficiency disease scurvy. Humans and primates lack the terminal enzyme in the biosynthetic pathway of ascorbic acid, l-gulonolactone oxidase, because the gene encoding for the enzyme has undergone substantial mutation so that no protein is produced.

ROLE IN HUMAN METABOLIC PROCESSES

- Vitamin C is an electron donor (reducing agent or antioxidant), and probably all of its biochemical and molecular functions can be accounted for by this function. The potentially protective role of vitamin C as an antioxidant.
- Vitamin C acts as an electron donor for 11 enzymes. Three of those enzymes are found in fungi but not in humans or other mammals. They are involved in

reutilisation pathways for pyrimidines and the deoxyribose moiety of deoxynucleosides. Of the 8 remaining human enzymes, three participate in collagen hydroxylation and two in carnitine biosynthesis; of the three enzymes which participate in collagen hydroxylation, one is necessary for biosynthesis of the catecholamine norepinephrine, one is necessary for amidation of peptide hormones, and one is involved in tyrosine metabolism.

- Ascorbate interacts with enzymes having either monooxygenase or dioxygenase activity. The monooxygenases dopamine b-monooxygenase and peptidyl-glycine a-monooxygenase incorporate a single oxygen atom into a substrate, either a dopamine or a glycine-terminating peptide. The remaining enzymes are dioxygenases which incorporate two oxygen atoms in two different ways. The enzyme 4-hydroxyphenylpyruvate dioxygenase incorporates two oxygen atoms into one product. The other dioxygenase incorporates one oxygen atom into succinate and one into the enzyme-specific substrate.
- The concentrations of vitamin C in gastric juice were several fold higher (median, 249 mmol/l; range, 43-909 mmol/l) than those found in the plasma of the same normal subjects (39 mmol/l, 14-101 mmol/l). Gastric juice vitamin C may prevent the formation of N-nitroso compounds, which are potentially mutagenic. High intakes of vitamin C correlate with reduced gastric cancer risk, but a cause-and-effect relationship has not been established. Vitamin C protects low-density lipoproteins ex vivo against oxidation and may function similarly in the blood.
- A common feature of vitamin C deficiency is anaemia. The antioxidant properties of vitamin C may stabilise folate in food and in plasma, and increased excretion of oxidised folate derivatives in human scurvy was

reported. Vitamin C promotes absorption of soluble non-haem iron possibly by chelation or simply by maintaining the iron in the reduced (ferrous, Fe^{2+}) form. The effect can be achieved with the amounts of vitamin C obtained in foods. However, the amount of dietary vitamin C required to increase iron absorption ranges from 25 mg upwards and depends largely on the amount of inhibitors, such as phytates and polyphenols, present in the meal.

OVERVIEW OF SIGNIFICANT SCIENTIFIC INFORMATION

- From the 15th century, scurvy was dreaded by seamen and explorers forced to subsist for months on diets of dried beef and biscuits. Scurvy was described by the Crusaders, during the sieges of numerous European cities, and as a result of the famine in 19th century Ireland. Three important manifestations of scurvy - gingival changes, pain in the extremities, and haemorrhagic manifestations - preceded oedema, ulcerations, and ultimately death. Skeletal and vascular lesions in scurvy probably arise from a failure of osteoid formation. In infantile scurvy the changes are mainly at the sites of most active bone growth; characteristic signs are a pseudoparalysis of the limbs caused by extreme pain on movement and caused by haemorrhages under the periosteum, as well as swelling and haemorrhages in areas of the gums surrounding erupting teeth. In adults one of the early, principle adverse effects of the collagen-related pathology may be impaired wound healing. Vitamin C deficiency can be detected from early signs of clinical deficiency, such as the follicular hyperkeratosis, petechial haemorrhages, swollen or bleeding gums, and joint pain, or from the very low concentrations of ascorbate in plasma, blood, or leukocytes. The Sheffield studies and later studies in Iowa were the first major attempts made to quantify vitamin C requirements. The studies indicated that the amount of vitamin C required to prevent or cure early signs of deficiency was between 6. 5 and 10 mg/day. This range represents the lowest physiologic requirement.

- Clinical signs of scurvy appear in men at intakes lower than 10 mg/day or when the whole body content falls below 300 mg. Such intakes are associated with plasma ascorbate concentrations below 11 mmol/l or leukocyte levels less than 2 nmol/ 10^8 cells. However, the plasma concentrations fall to around 11 mmol/l when dietary vitamin C is between 10 and 20 mg/day. At intakes greater than 25-35 mg/ day, plasma concentrations start to rise steeply, indicating a greater availability of vitamin C for metabolic needs. In general, plasma ascorbate closely reflects the dietary intake and ranges between 20 and 80 mmol/l. Note that during infection or physical trauma, an increase in the number of circulating leukocytes occurs and these take up vitamin C from the plasma. Therefore, both plasma and leukocyte levels may not be very precise indicators of body content or status at such times. However, leukocyte ascorbate remains a better indicator of vitamin C status than plasma ascorbate most of the time and only in the period immediately after the onset of an infection are both values unreliable.
- Intestinal absorption of vitamin C is by an active, sodium-dependent, energy-requiring, carrier-mediated transport mechanism and as intakes increase, the tissues progressively become more saturated. The physiologically efficient, renal-tubular reabsorption mechanism retains vitamin C in the tissues up to a whole body content of ascorbate of about 20 mg/kg body weight. However, under steady state conditions, as intakes rise from around 100 mg/ day there is an increase in urinary output in so that

at 1000 mg/day almost all absorbed vitamin C is excreted.

DEFINITION OF POPULATION AT RISK

- The populations at risk of vitamin C deficiency are those for whom the fruit and vegetable supply is minimal. Epidemics of scurvy are associated with famine and war, when people are forced to become refugees and food supply is small and irregular. Persons in whom the total body vitamin C content is saturated can subsist without vitamin C for approximately 2 months before the appearance of clinical signs, and as little as 6. 5-10 mg/day vitamin C will prevent the appearance of scurvy. In general, vitamin C status will reflect the regularity of fruit and vegetable consumption but also socio-economic conditions, because intake is determined not just by availability, but by cultural preferences and cost.
- In Europe and the United States an adequate intake of vitamin C is indicated by the results of various national surveys. In the United Kingdom and Germany, the mean dietary intakes of vitamin C in adult men and women were 87 and 76 and 75 and 72 mg/day, respectively. In addition, a recent survey of elderly men and women in the United Kingdom reported vitamin C intakes of 72 (SD 61) and 68 (SD 60) mg/day, respectively. In the United States, in the National Health and Nutrition Examination Survey, the median consumption of vitamin C from foods during the years 1988-91 was 73 and 84 mg/day in men and women, respectively. In all these studies there was a wide variation in vitamin C intake and 25-30 percent of the US population consumed less than 2. 5 servings of fruit and vegetables daily. Likewise a survey of Latin American children in the United States suggested that less than 15 percent consumed the recommended intake of fruits and vegetables. It is not possible to relate servings of

fruits and vegetables to an exact amount of vitamin C, but the World Health Organization (WHO) dietary goal of 400 g aimed to provide sufficient vitamin C to meet the 1970 Food and Agriculture Organization of the United Nations (FAO)/WHO guidelines - that is, approximately 20-30 mg/day - and lower the risk of chronic disease. The WHO goal has been roughly translated into the recommendation of five portions per day.

- Reports from India show that the available supply of vitamin C is 43 mg/capita/day, and in the different states of India it ranges from 27 to 66 mg/day. In one study, low-income children consumed as little as 8. 2 mg/day of vitamin C in contrast to a well-to-do group of children where the intake was 35. 4 mg/day. Other studies done in developing countries found plasma vitamin C concentrations lower than those reported for developed countries, for example, 20-27 mmol/l for apparently healthy adolescent boys and girls in China and 3-54 mmol/l (median 14 mmol/l) for similarly aged Gambian nurses, although values obtained in a group of adults from. However, it is difficult to assess the extent to which sub-clinical infections are lowering the plasma vitamin C concentrations seen in such countries.
- Data describing a positive association between vitamin C consumption and health status are frequently reported, but intervention studies do not support the observations. Low plasma concentrations are reported in patients with diabetes and infections and in smokers, but the relative contribution of diet and stress to these situations is uncertain. Epidemiologic studies indicate that diets with a high vitamin C content have been associated with lower cancer risk, especially for cancers of the oral cavity, oesophagus, stomach, colon, and lung. However, there appears to be no effect of consumption of vitamin C supplements on the

development of colourectal adenoma and stomach cancer (52-54), and data on the effect of vitamin C supplementation on coronary heart disease and cataract development are conflicting (55-74). Currently there is no consistent evidence from population studies that heart disease, cancers, or cataract development are specifically associated with vitamin C status. This of course does not preclude the possibility that other components in vitamin C - rich fruits and vegetables provide health benefits, but it is not yet possible to separate such an effect from other factors such as lifestyle patterns of people who have a high vitamin C intake.

DIETARY SOURCES OF VITAMIN C AND LIMITATIONS TO VITAMIN C

- Ascorbate is found in many fruits and vegetables. Citrus fruits and juices are particularly rich sources of vitamin C but other fruits including cantaloupe, honeydew melon, cherries, kiwi fruits, mangoes, papaya, strawberries, tangelo, watermelon, and tomatoes also contain variable amounts of vitamin C. Vegetables such as cabbage, broccoli, Brussels sprouts, bean sprouts, cauliflower, kale, mustard greens, red and green peppers, peas, tomatoes, and potatoes may be more important sources of vitamin C than fruits. This is particularly true because the vegetable supply often extends for longer periods during the year than does the fruit supply.
- In many developing countries, limitations in the supply of vitamin C are often determined by seasonal factors (i. e., the availability of water, time, and labour for the management of household gardens and the short harvesting season of many fruits). For example, mean monthly ascorbate intakes ranged from 0 to 115 mg/day in one Gambian community in which peak intakes coincided with the seasonal duration of the mango crop and to a lesser extent

with orange and grapefruit harvests. These fluctuations in dietary ascorbate intake were closely reflected by corresponding variations in plasma ascorbate (11. 4-68. 4 mmol/L) and human milk ascorbate (143-342 mmol/L).

- Vitamin C is also very labile, and the loss of vitamin C on boiling milk provides one dramatic example of a cause of infantile scurvy. The vitamin C content of food is strongly influenced by season, transport to market, shelf life, time of storage, cooking practices, and chlorination of water. Cutting or bruising of produce releases ascorbate oxidase. Blanching techniques inactivate the oxidase enzyme and help to preserve ascorbate as also will low pH, as in the preparation of sauerkraut. In contrast, heating and exposure to copper or iron or to mildly alkaline conditions destroys the vitamin, and too much water can leach it from the tissues during cooking.
- The use of citrus fruits by the British navy in the 18th century gave rise to the term 'limey', a colloquial term for British sailors. However, it is important to realise that the amount of vitamin C in a food is usually not the major determinant of a food's importance for supply, but rather regularity of intake. For example, in countries where the potato is an important staple food and refrigeration facilities are limited, seasonal variations in plasma ascorbate are due to the considerable deterioration in the potato's vitamin C content during storage; the content can decrease from 30 to 8 mg/100 g over 8-9 months. Such data can indicate the important contribution the potato can make to human vitamin C requirements even though the potato vitamin C concentration is low.
- An extensive study has been made of losses of vitamin C during the packaging, storage, and

cooking of blended foods. Data from a US Agency for Internation Development programme show that vitamin C losses from packaging and storage in polythene bags of such relief foods are much less significant than the 52-82 percent losses attributable to conventional cooking procedures.

INFORMATION USED TO DERIVE DIETARY REQUIREMENT OF VITAMIN C

- At saturation the whole body content of ascorbate in adult males is approximately 20 mg/kg, or 1500 mg. Clinical signs of scurvy appear when the whole body content falls below 300-400 mg, and the last signs disappear when the body content reaches about 1000 mg. In these experiments, ascorbate in the whole body was catabolised at an approximate rate of 3 percent/day (2. 9 percent/day, SD 0. 6).

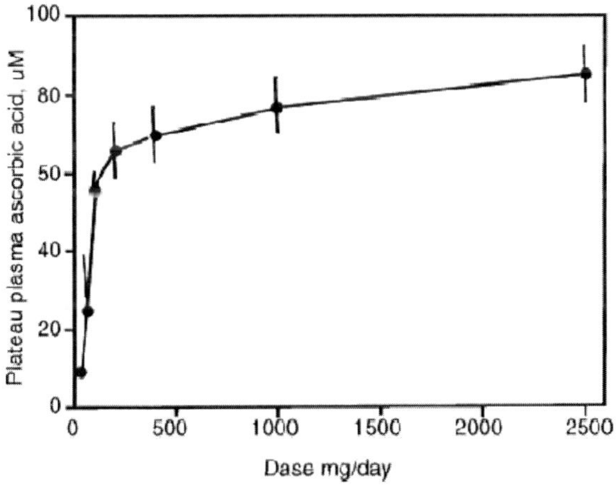

Fig. Relationship between intake and plasma concentrations of vitamin C

- There is a sigmoidal relationship between intake and plasma concentrations of vitamin C. Below 30 mg/ day, plasma concentrations are around 11 mmol/l. Above this intake, plasma concentrations increase steeply to 60 mmol/l and plateau at around 80 mmol/ l, which represents the renal threshold. Under near steady state conditions, plateau concentrations of vitamin C are achieved by intakes in excess of 200 mg/day. At low doses dietary vitamin C is almost completely absorbed, but over the range of usual dietary intakes (30-180 mg/day), absorption may decrease to 75 percent because of competing factors in the food.

Vitamin C Toxicity

- The potential toxicity of excessive doses of supplemental vitamin C relates to intra-intestinal events and to the effects of metabolites in the urinary system. Intakes of 2-3 g/day of vitamin C produce unpleasant diarrhoea from the osmotic effects of the unabsorbed vitamin in the intestinal lumen in most people. Gastrointestinal disturbances can occur after ingestion of as little as 1 g because approximately half of the amount would not be absorbed at this dose.
- Oxalate is an end product of ascorbate catabolism and plays an important role in kidney stone formation. Excessive daily amounts of vitamin C produce hyperoxaluria. In four volunteers who received vitamin C in the range of 5-10 g/day, this amounted to approximately a doubling of urinary oxalate excretion, from 50 to 87 mg/day (range 60-126 mg/day). However, the risk of oxalate stones formation may become significant at high intakes of vitamin C (>1 g), particularly in subjects with high amounts of urinary calcium.
- Vitamin C may precipitate haemolysis in some people, including those with glucose-6-phosphate dehydrogenase deficiency, paroxysmal nocturnal haemaglobinuria, or other conditions where increased risk of red cell haemolysis may occur or where protection against the removal of the products of iron metabolism may be impaired, as in people with the haptoglobin Hp2-2 phenotype. Of these

conditions, only the haptoglobin Hp 2-2 condition was associated with abnormal vitamin C metabolism (lower plasma ascorbate than expected) under conditions where intake of vitamin C was provided mainly from dietary sources. Therefore, 1 g vitamin C appears to be the advisable upper limit of dietary intake.

VITAMIN D

THE ROLE OF VITAMIN D IN HUMAN

Vitamin D is required to maintain normal blood levels of calcium and phosphate, that are in turn needed for the normal mineralisation of bone, muscle contraction, nerve conduction, and general cellular function in all cells of the body. Vitamin D achieves this after its conversion to the active form 1,25-dihydroxyvitamin D [1,25-$(OH)_2$D], or calcitriol. This active form regulates the transcription of a number of vitamin D-dependent genes coding for calcium-transporting proteins and bone matrix proteins.

Vitamin D also modulates the transcription of cell cycle proteins, that decrease cell proliteration and increase cell differentiation of a number of specialised cells of the body (e. g., osteoclastic precursors, enterocytes, keratinocytes, etc.). This property may explain the actions of vitamin D in bone resorption, intestinal calcium transport, and skin. Vitamin D also possesses immuno-modulatory properties that may alter responses to infections in vivo. The cell differentiating and immuno-modulatory properties underlie the reason why vitamin D derivatives are now used successfully in the treatment of psoriasis and other skin disorders.

Clinical assays measure 1,25-$(OH)_2D_2$ and 1,25-$(OH)_2D_3$, collectively called 1,25-$(OH)_2$D. Similarly, calcidiol is measured as 25-OH-D but it is a mixture of 25-OH-D_2 and 25-OH-D_3. For the purposes of this document, 1,25-$(OH)_2$D and 25-OH-D will be used to refer to calcitriol and calcidiol, respectively.

THE ROLE OF VITAMIN D

Vitamin D, a seco-steroid, can either be made in the skin from a cholesterol-like precursor (7-dehydrocholesterol) by exposure to sunlight or can be provided pre-formed in the diet. The version made in the skin is referred to as vitamin D_3 whereas the dietary form can be vitamin D_3 or a closely related molecule of plant origin known as vitamin D_2. Because vitamin D can be made in the skin, it should not strictly be called a vitamin, and some nutritional texts refer to the substance as a prohormone and to the two forms as cholecalciferol (D_3) or ergocalciferol (D_2).

From a nutritional perspective, the two forms are metabolised similarly in humans, are equal in potency, and can be considered equivalent. It is now firmly established that vitamin D_3 is metabolised first in the liver to 25-hydroxyvitamin-D (25-OH-D or calcidiol) and subsequently in the kidneys to 1,25-$(OH)_2$D to produce a biologically active hormone. The 1,25-$(OH)_2$D, like all vitamin D metabolites, is present in the blood complexed to vitamin D binding protein, a specific a-globulin. The 1,25-$(OH)_2$D is believed to act on target cells similarly to the way a steroid hormone would act. Free hormone crosses the plasma membrane and interacts with a specific nuclear receptor known as the vitamin D receptor, a DNA-binding, zinc-finger protein with a molecular weight of 55,000. This ligand-receptor complex binds to a specific vitamin D-responsive element and, with associated transcription factors (e. g., retinoid X receptor), enhances transcription of mRNAs which code for calcium-transporting proteins, bone matrix proteins, or cell cycle-regulating proteins. As a result of these processes,1,25-$(OH)_2$D stimulates intestinal absorption of calcium and phosphate and mobilises calcium and phosphate by stimulating bone resorption. These functions serve the common purpose of restoring blood levels of calcium and phosphate to normal when concentrations of the two ions are low.

Lately, interest has focused on other cellular actions of 1,25-$(OH)_2$D. With the discovery of 1,25-$(OH)_2$D receptors in many classical non-target tissues such as brain, various bone marrow-derived cells, skin, thymus, etc., the view has been expressed that 1,25-$(OH)_2$D induces fusion and differentiation of macrophages. This effect has been widely interpreted to mean that the natural role of 1,25-$(OH)_2$D is to induce osteoclastogenesis from colony forming units-granulatory monocytes in the bone marrow.

The 1,25-$(OH)_2$D also suppresses interleukin 2 production in activated T lymphocytes, an effect which suggests the hormone might play a role in immuno-modulation in vivo. Other tissues (e. g., skin) are directly affected by exogenous administration of vitamin D, though the physiologic significance of these effects is poorly understood. The pharmacologic effects of 1,25-$(OH)_2$D are profound and have resulted in the development of vitamin D analogues, that are approved for use in hyper-proliferative conditions such as psoriasis.

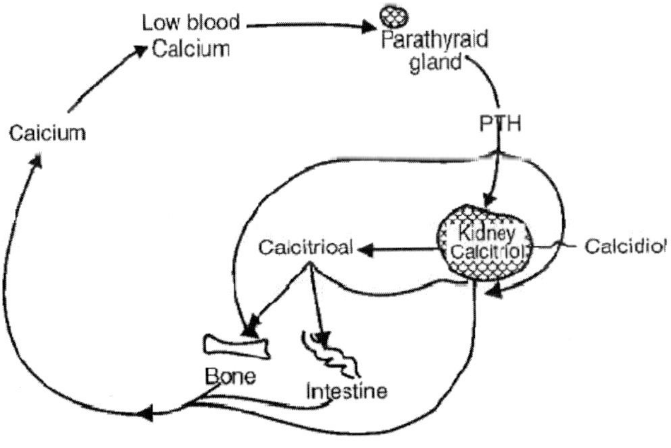

Fig. Calcium homeostasis

In calcium homeostasis 1,25-$(OH)_2$D works in conjunction with parathyroid hormone (PTH) to produce its beneficial effects on the plasma levels of ionised calcium and phosphate. The physiologic loop starts with calcium sensing by the calcium receptor of the parathyroid gland. When the level of ionised calcium in plasma falls, PTH is secreted by the parathyroid gland and stimulates the tightly regulated renal enzyme 25-OH-D-1-aa-hydroxylase to make more 1,25-$(OH)_2D$ from the large circulating pool of 25-OH-D.

The resulting increase in 1,25-$(OH)_2D$ (with the rise in PTH) causes an increase in calcium transport within the intestine, bone, and kidney. All these events raise plasma calcium levels back to normal, that in turn is sensed by the calcium receptor of the parathyroid gland. The further secretion of PTH is turned off not only by the feedback action of calcium, but also by a short feedback loop involving 1,25-$(OH)_2D$ directly suppressing PTH synthesis in the parathyroid gland.

Although this model oversimplifies the events involved in calcium homeostasis it is easy to see from it that sufficient 25-OH-D must be available to provide adequate 1,25-$(OH)_2D$ synthesis and hence an adequate level of plasma calcium and that vitamin D deficiency will result in inadequate 25-OH-D and 1,25-$(OH)_2D$ synthesis, inadequate calcium homeostasis, and a constantly elevated PTH level.

It becomes evident from this method of presentation of the role of vitamin D that the nutritionist can focus on the plasma levels of 25-OH-D and PTH to gain an insight into vitamin D status. Not shown but also important is the endpoint of the physiologic action of vitamin D, namely adequate plasma calcium and phosphate ions, that provide the raw materials for bone mineralisation.

POPULATIONS AT RISK FOR VITAMIN D DEFICIENCY

Infants

Infants constitute a population at risk for vitamin D deficiency because of relatively large vitamin D needs brought about by their high rate of skeletal growth. At birth, infants have acquired in utero the vitamin D stores that must carry them through the first months of life. A recent survey of French neonates revealed that 64 percent had 25-OH-D values below 30 nmol/l, the lower limit of the normal range. Breast-fed infants are particularly at risk because of the low concentrations of vitamin D in human milk. This problem is further compounded in some infants fed human milk by a restriction in exposure to ultraviolet (UV) light for seasonal, latitudinal, cultural, or social reasons. Infants born in the autumn months at extremes of latitude are particularly at risk because they spend the first 6 months of their life indoors and therefore have little opportunity to synthesise vitamin D in their skin during this period. Consequently, although vitamin D deficiency is rare in developed countries, sporadic cases of rickets are still being reported in many northern cities but almost always in infants fed human milk.

Infant formulas are supplemented with vitamin D at levels ranging from 40 international units (IUs) or 1 mg /418. 4 kJ to 100 IU or 2. 5 mg /418. 4 kJ, that provide approximately between 6 mg and 15 mg of vitamin D, respectively. These amounts of dietary vitamin D are sufficient to prevent rickets.

Adolescents

Another period of rapid growth of the skeleton occurs at puberty and increases the need not for the vitamin D itself, but for the active form 1,25-$(OH)_2D$. This need results from the increased conversion of 25-OH-D to 1,25-$(OH)_2D$ in adolescents. Furthermore, unlike infants, adolescents are usually outdoors and therefore usually are exposed to UV light sufficient for synthesising vitamin D for their needs. Excess production of vitamin D in the summer and early fall months is stored mainly in the adipose tissue and is available to sustain high growth rates in the winter months that follow. Insufficient vitamin D stores during these periods of increased growth can lead to vitamin D insufficiency.

Elderly

Over the past 20 years, clinical research studies of the basic biochemical machinery handling vitamin D have suggested an age-related decline in many key steps of vitamin D action including rate of skin synthesis, rate of hydroxylation leading to activation to the hormonal form, and response of target tissues (e. g., bone) as well as reduced skin exposure. Not surprisingly a number of independent studies from around the world have shown that there appears to be vitamin D deficiency in a subset of the elderly population, as characterised by low blood levels of 25-OH-D coupled with elevations of plasma PTH and alkaline phosphatase.

There is evidence that this vitamin D deficiency contributes to declining bone mass and increases the incidence of hip fractures. Although some of these studies may exaggerate the extent of the problem by focusing on institutionalised individuals or in-patients with decreased sun exposures, in general they have forced health professionals to re-address the intakes of this segment of society and look at potential solutions to correct the problem. Several groups have found that modest increases in vitamin D intakes (between 10 and 20 μg/day) reduce the rate of bone loss and the fracture rate.

These findings have led agencies and researchers to suggest an increase in recommended vitamin D intakes for the elderly from the suggested 2. 5-5 μg /day to a value that is able to maintain normal 25-OH-D levels in the elderly, such as 10-15 μg/day. This vitamin D intake results in lower rates of bone loss and is suggested for the middle-aged (50-70 years) and old-aged (>70 years) populations. The increased requirements are justified mainly on the grounds of the reduction in skin synthesis of vitamin D, a linear reduction occurring in both men and women, that begins with the thinning of the skin at age 20 years.

Pregnancy and Lactation

Elucidation of the changes in calciotropic hormones occurring during pregnancy and lactation has revealed a role for vitamin D in the former but probably not the latter. Even in pregnancy, the changes in vitamin D metabolism which occur, namely an increase in the maternal plasma levels of 1,25-$(OH)_2$D due to a putative placental synthesis of the hormone, do not seem to impinge greatly on the maternal vitamin D requirements. The concern that modest vitamin D supplementation might be deleterious to the foetus is not justified. Furthermore, because transfer of vitamin D from mother to foetus is important for establishing the newborn's growth rate, the goal of ensuring adequate vitamin D status with conventional prenatal vitamin D supplements probably should not be discouraged.

In lactating women there appears to be no direct role for vitamin D because increased calcium needs are regulated by PTH-related peptide, and recent studies have failed to show any change in vitamin D metabolites during lactation. As stated above, the vitamin D content of human milk is low. Consequently, there is no great drain on maternal vitamin D reserves either to regulate calcium homeostasis or to supply the need of human milk.

Because human milk is a poor source of vitamin D, rare cases of nutritional rickets are still found, but these are almost always in breast-fed babies deprived of sunlight exposure. Furthermore, there is little evidence that increasing calcium or vitamin D supplements to lactating mothers results in an increased transfer of calcium or vitamin D in milk. Thus, the current thinking, based on a clearer understanding of the role of vitamin D in lactation, is that there is little purpose in recommending additional vitamin D for lactating women. The goal for mothers who breast-feed their infants seems to be merely to ensure good nutrition and sunshine exposure in order to ensure normal vitamin D status during the perinatal period.

EVIDENCE USED FOR ESTIMATING RECOMMENDED VITAMIN D INTAKE

Lack of Accuracy in Estimating Dietary Intake and Skin Synthesis

The unique problem of estimating total intake of a substance that can be provided in the diet or made in the skin by exposure to sunlight makes it difficult to estimate adequate total intakes of vitamin D for the general population. Accurate food composition data are not available for vitamin D, accentuating the difficulty for estimating dietary intakes. Whereas this has led two recent US national surveys to avoid attempting this task, the second National Health and Nutrition Examination Survey (NHANES II) estimated vitamin D intakes to be 2. 9 μg/day and 2. 3 μg/day for younger and older women, respectively. A recent study of elderly women concurred with this assessment, finding an intake of 3. 53 mg/day.

Skin synthesis is equally difficult to estimate, being affected by such imponderables as age, season, latitude, time of day, skin exposure, sun screen use, etc. In vitamin D - replete individuals, estimates of skin synthesis are put at around 10 μg /day, with total intakes estimated at 15 μg/day.

Use of Plasma 25-OH-D as a Measure of Vitamin D Status

Numerous recent studies have used plasma 25-OH-D as a measure of vitamin D status, and there is a strong presumptive relationship of this variable with bone status. Thus, it is not surprising that several nutritional committees (e. g., the Food and Nutrition Board of the US National Academy of Sciences' Institute of Medicine in conjunction with Health Canada) have chosen to use a biochemical basis for estimating required intakes and used these estimates to derive recommended intakes.

The method used involved the estimation of the mean group dietary intake of vitamin D required to maintain the plasma 25-OH-D levels above 27 nmol/l, that is necessary to ensure normal bone health. Previously, many studies had established 27 nmol/l as the lower limit of the normal range. This dietary intake of vitamin D for each population group was rounded to the nearest 50 IU (1. 25 μg) and then doubled to cover the needs of all individuals within that group irrespective of sunlight exposure. They termed this amount adequate intake (AI) and used it in place of recommended dietary allowance (RDA), that had been used by US agencies since 1941. We are recommending the use of those figures here as recommended nutrient intakes (RNIs) because it is an entirely logical approach to estimating the vitamin D needs for the whole population.

Total number of samples used in data analysis: 18,323; mean 25-OH-D value for United States: 25. 89 ng/ml (+/- 11. 08). Frequency distribution of serum or plasma 25-OH-D. Values for all ages, ethnicity groups, both sexes. [b]High values: four values 90-98 ng/ml, one value of 160. 3 ng/ml. Values < 5 ng/ml (lowest standard) entered arbitrarily in the database as "3"..

Because study after study had been recommending increases in vitamin D intakes for the elderly, it might have been expected that the proposed increases in suggested intakes from 5 μg/day (RDAs in the United States [43], RNIs in Canada [44]) to 10 μg/day or 15 μg/day (AI) would be welcomed. However, a recent editorial in a prominent medical journal attacked the recommendations as being too conservative. Of course, such in-patients are by definition sick and should not be used to calculate normal intakes. Indeed, the new NHANES III study of 18 323 normal individuals from all regions of the United States suggests that approximately 5 percent had values of 25-OH-D below 27 nmol/l. Although the data are skewed by sampling biases that favour sample collection in the southern states in winter months and northern states in the summer months, even subsets of data collected in northern states in September give the incidence of low 25-OH-D in the elderly in the 6-18 percent range as compared with 57 percent in the institutionalised in-patient population.

Ideally, such measurements of the normal population should be made at the end of the winter months and before UV irradiation has reached a strength sufficient to allow skin synthesis of vitamin D. Thus, the NHANES III study may still underestimate the incidence of hypovitaminosis D in a northern elderly population in winter. Nevertheless, in lieu of additional studies of selected human populations, it would seem that the recommendations of the Food and Nutrition Board are reasonable guidelines for vitamin D intakes, at least for the near future. This considered approach allows for a period of time to monitor the potential shortfalls of the new recommendations as well as to assess whether the suggested guidelines can be attained, a point that was repeatedly stated about the RDAs.

INTAKES FOR VITAMIN D

In recommending intakes for vitamin D, it must be recognised that in most locations in the world in a broad band around the equator (between latitudes 42°N and 42°S), the most physiologically relevant and efficient way of acquiring vitamin D is to synthesise it endogenously in the skin from 7-dehydrocholesterol by sun (UV) light exposure. In most situations, approximately 30 minutes of skin exposure (without sunscreen) of the arms and face to sunlight can provide all the daily vitamin D needs of the body. However, skin synthesis of vitamin D is negatively influenced by factors which may reduce the ability of the skin to provide the total needs of the individual:

- *Latitude and season:* Both influence the amount of UV light reaching the skin;
- *The ageing process:* Thinning of the skin reduces the efficiency of this synthetic process;
- *Skin pigmentation:* The presence in the skin of darker pigments interferes with UV light reaching the appropriate layer of the skin;
- *Clothing:* Virtually complete covering of the skin for medical, social, cultural, or religious reasons leaves insufficient skin exposed to sunlight; and
- *Sunscreen use:* Widespread and liberal use of sun-blockers reduces skin damage by the sun but also deleteriously affects synthesis of vitamin D.

Because not all of these problems can be solved in all geographic locations, particularly during winter at latitudes higher than 42° where synthesis is virtually sero, it is recommended that individuals not synthesising vitamin D should correct their vitamin D status by consuming the amounts of vitamin D appropriate for their age group.

Table: RNIs for vitamin D according to age groups

Age group	RNI μg/day
Infants	
0-6 months	5
7-12 onths	5
1-3 years	5
4-6 years	5
7-9 years	5
Adolescents,	
10-18 years	5
Adults	
19-50 years	5
Older adults, 51-65 years	10
Elderly adults, 65+ years	15
Pregnant women	5
Lactating women	5

Vitamin D toxicity

The adverse effects of high vitamin D intakes - hypercalciuria and hypercalcemia - do not occur at these new recommended intake levels. In fact, it is worth noting that the recommended intakes for all age groups are still well below the lowest observed adverse effect level of 50 μg/day and have not yet even reached the no observed adverse effect level of 20 μg/day. Outbreaks of idiopathic infantile hypercalcemia in the United Kingdom in the post-World War II era led to the withdrawal of vitamin D fortification from all foods in that country because of concerns that they were due to hypervitaminosis D. There are some suggestions in the literature that these outbreaks of idiopathic infantile hypercalcemia may have been multifactorial with genetic and dietary components and were not just due to technical problems with over-fortification as was assumed. In retrospect, the termination of the vitamin D fortification may have been counter productive because it exposed segments of the UK community to vitamin deficiency and may have discouraged other nations from starting vitamin D fortification programmes. This is all the more cause for concern because hypovitaminosis D is still a problem worldwide, particularly in developing countries at high latitudes and in countries where skin exposure to sunlight is discouraged.

VITAMIN E

THE ROLE OF VITAMIN E IN HUMAN

A large body of scientific evidence indicates that reactive free radicals are involved in many diseases, including heart disease and cancers. Cells contain many potentially oxidizable substrates such as polyunsaturated fatty acids (PUFAs), proteins, and DNA. Therefore, a complex antioxidant defence system normally protects cells from the injurious effects of endogenously produced free radicals as well as from species of exogenous origin such as cigarette smoke and pollutants. Should our exposure to free radicals exceed the protective capacity of the antioxidant defence system, a phenomenon often referred to as oxidative stress, then damage to biologic molecules may occur.

There is considerable evidence that disease causes an increase in oxidative stress; therefore, consumption of foods rich in antioxidants, which are potentially able to quench or neutralise excess radicals, may play an important role in modifying the development of such diseases.

Vitamin E is the major lipid-soluble antioxidant in the cell antioxidant defence system and is exclusively obtained from the diet. The term "vitamin E" refers to a family of eight naturally occurring homologues that are synthesised by plants from homogentisic acid. All are derivatives of 6-chromanol and differ in the number and position of methyl groups on the ring structure. The four tocopherol homologues (d-a-, d-b-, d-g-, and d-d-) have a saturated 16-carbon phytyl side chain, whereas the tocotrienols homologues (d-a-, d-b-, d-g-, and d-d-) have three double bonds on the side chain.

There is also a widely available synthetic form, dl-a-tocopherol, prepared by coupling trimethylhydroquinone with isophytol. This consists of a mixture of eight stereoisomers in approximately equal amounts; these isomers are differentiated by rotations of the phytyl chain in various directions that do not occur naturally.

For dietary purposes, vitamin E activity is expressed as a-tocopherol equivalents (a-TEs). One a-TE is the activity of 1 mg RRR-a-tocopherol (d-a-tocopherol). To estimate the a-TE of mixed diet containing natural forms of vitamin E, the number of milligrams of b-tocopherol should be multiplied by 0. 5, g-tocopherol by 0. 1, and a-tocotrienol by 0. 3. Any of the synthetic all-rac-a-tocopherol (dl-a-tocopherol) should be multiplied by 0. 74. One milligram of the latter compound in the acetate form is equivalent to 1 IU of vitamin E.

Vitamin E is an example of a phenolic antioxidant. Such molecules readily donate the hydrogen from the hydroxyl (-OH) group on the ring structure to free radicals, which then become unreactive. On donating the hydrogen, the phenolic compound itself becomes a relatively unreactive free radical because the unpaired electron on the oxygen atom is usually delocalised into the aromatic ring structure thereby increasing its stability.

The major biologic role of vitamin E is to protect PUFAs and other components of cell membranes and low-density lipoprotein (LDL) from oxidation by free radicals. Vitamin E is located primarily within the phospholipid bilayer of cell membranes. It is particularly effective in preventing lipid peroxidation, a series of chemical reactions involving the oxidative deterioration of PUFAs. Elevated levels of lipid peroxidation products are associated with numerous diseases and clinical conditions. Although vitamin E is primarily located in cell and organelle membranes where it can exert its maximum protective effect, its concentration may only be one molecule for every 2000 phospholipid molecules. This suggests that after its reaction with free radicals it is rapidly regenerated, possibly by other antioxidants.

Absorption of vitamin E from the intestine depends on adequate pancreatic function, biliary secretion, and micelle formation. Conditions for absorption are like those for dietary lipid, that is, efficient emulsification, solubilisation within mixed bile salt micelles, uptake by enterocytes, and secretion into the circulation via the lymphatic system. Emulsification takes place initially in the stomach and then in the small intestine in the presence of pancreatic and biliary secretions.

The resulting mixed micelle aggregates the vitamin E molecules, solubilises the vitamin E, and then transports it to the brush border membrane of the enterocyte probably by passive diffusion. Within the enterocyte, tocopherol is incorporated into chylomicrons and secreted into the intracellular space and lymphatic system and subsequently into the blood stream. Tocopherol esters, present in processed foods and vitamin supplements, must be hydrolysed in the small intestine before absorption.

Vitamin E is transported in the blood by the plasma lipoproteins and erythrocytes. Chylomicrons carry tocopherol from the enterocyte to the liver, where they are incorporated into parenchymal cells as chylomicron remnants. The catabolism of chylomicrons takes place in the systemic circulation through the action of cellular lipoprotein lipase. During this process tocopherol can be transferred to high-density lipoproteins (HDLs).

The tocopherol in HDLs can transfer to other circulating lipoproteins, such as LDLs and very low-density lipoproteins (VLDLs). During the conversion of VLDL to LDL in the circulation, some a-tocopherol remains within the core lipids and thus is incorporated in LDL. Most a-tocopherol then enters the cells of peripheral tissues within the intact lipoprotein through the LDL receptor pathway, although some may be taken up by membrane binding sites recognising apolipoprotein A-I and A-II present on HDL.

Although the process of absorption of all the tocopherol homologues in our diet is similar, the a form predominates in blood and tissue. This is due to the action of binding proteins that preferentially select the a form over the others. In the first instance, a 30-kDa binding protein unique to the liver cytoplasm preferentially incorporates a-tocopherol in the nascent VLDL. This form also accumulates in non-hepatic tissues, particularly at sites where free radical production is greatest, such as in the membranes of mitochondria and endoplasmic reticulum in the heart and lungs.

Hepatic intracellular transport may be expedited by a 14. 2-kDa binding protein that binds a-tocopherol in preference to the other homologues. Other proteinaceous sites with apparent tocopherol-binding abilities have been found on erythrocytes, adrenal membranes, and smooth muscle cells. These may serve as vitamin E receptors which orient the molecule within the membrane for optimum antioxidant function.

These selective mechanisms explain why vitamin E homologues have markedly differing antioxidant abilities in biologic systems and illustrates the important distinction between the in vitro antioxidant effectiveness of a substance in the stabilisation of, for example, a food product and its in vivo potency as an antioxidant. From a nutritional perspective, the most important form of vitamin E is a-tocopherol; this is corroborated in animal model tests of biopotency which assess the ability of the various homologues to prevent foetal absorption and muscular dystrophies.

Plasma vitamin E concentrations vary little over a wide range of dietary intakes. Even daily supplements of the order of 1600 IU/day for 3 weeks only increased plasma levels 2-3 times and on cessation of treatment plasma levels returned to pretreatment levels in 5 days. Likewise, tissue concentrations only increased by a similar amount when patients undergoing heart surgery were given 300 mg/day of the natural stereoisomer for 2 weeks preoperatively. Kinetic studies with deuterated tocopherol suggest that there is rapid equilibration of new tocopherol in erythrocytes, liver, and spleen but that turnover in other tissues such as heart, muscle, and adipose tissue is much slower. The brain is markedly resistant to depletion and repletion with vitamin E. This presumably reflects an adaptive mechanism to avoid detrimental oxidative reactions in this key organ.

The primary oxidation product of a-tocopherol is a tocopheryl quinone that can be conjugated to yield the glucuronate after prior reduction to the hydroquinone. This is excreted in the bile or further degraded in the kidneys to a-tocopheronic acid and hence excreted in the bile. Those vitamin E homologues not preferentially selected by the hepatic binding proteins are eliminated during the process of nascent VLDL secretion in the liver and probably excreted via the bile. Some vitamin E may also be excreted via skin sebaceous glands.

Table: Approximate biological activity of naturally occurring tocopherols and tocotrienols compared with d-a-tocopherol

Common name	Biological activity compared with d-a-tocopherol, %
d-a-tocopherol	100
d-b-tocopherol	50
d-g-tocopherol	10
d-d-tocopherol	3
d-a-tocotrienol	30
d-b-tocotrienol	5
d-g-tocotrienol	not known
d-d-tocotrienol	not known

Defining Populations at Risk of Vitamin E Deficiency

There are many signs of vitamin E deficiency in animals most of which are related to damage to cell membranes and leakage of cell contents to external fluids. Disorders provoked, for example, by traces of peroxidised PUFAs in the diets of animals with low vitamin E status are cardiac or skeletal myopathies, neuropathies, and liver necrosis. Muscle and neurological problems are also a consequence of human vitamin E deficiency. Early diagnostic signs of deficiency include leakage of muscle enzymes such as creatine kinase and pyruvate kinase into plasma, increased levels of lipid peroxidation products in plasma, and increased erythrocyte haemolysis.

The assessment of the vitamin E requirement for humans is confounded by the infrequent occurrence of clinical signs of deficiency because these usually only develop in adults with fat-malabsorption syndromes or liver disease, in individuals with genetic anomalies in transport or binding proteins, and possibly in premature infants. This suggests that diets contain sufficient vitamin E to satisfy nutritional needs.

Several animal models suggest that increasing intakes of vitamin E inhibit the progression of vascular disease by preventing the oxidation of LDL. Evidence suggests that oxidised lipoprotein is a key event in the development of the atheromatous plaque which may ultimately occlude the blood vessel.

Table: Diseases and syndromes in animals associated with vitamin E deficiency and excess intakes of polyunsaturated fatty acids

Syndrome	Affected organ or tissue	Species
Encephalomalacia	Cerebellum	Chick
Exudative diathesis	Vascular	Turkey
Microcytic anaemia	Blood, bone marrow	Chick
Macrocytic anaemia	Blood, bone marrow	Monkey
Pancreatic fibrosis	Pancreas	Chick, mouse
Liver necrosis	Liver	Pig, rat
Muscular degeneration	Skeletal muscle	Pig, rat, mouse
Microangiopathy	Heart muscle	Pig, lamb, calf
Kidney degeneration	Kidney tubules	Monkey, rat

Steatitis	Adipose tissue	Pig, chick
Testicular degeneration	Testes	Pig, calf, chick
Malignant hyperthermia	Skeletal muscle	Pig

Human studies, however, have been less consistent in providing evidence for a role of vitamin E in preventing heart disease. Vitamin E supplements reduce ex vivo oxidizability of plasma LDLs but there is no correlation between ex vivo lipoprotein oxidizability and endogenous vitamin E levels in an unsupplemented population. Likewise, the few randomised double-blind, placebo-controlled intervention trials with human volunteers which focused on the relationship between vitamin E and cardiovascular disease have given inconsistent results. There was a marked reduction in non-fatal myocardial infarction in patients with coronary artery disease (as defined by angiogram) who were randomly assigned to take pharmacologic doses of vitamin E (400 and 800 mg/day) or placebo in the Cambridge Heart Antioxidant Study involving 2000 men and women. However, the incidence of major coronary events in male smokers who received 20 mg/day of vitamin E for approximately 6 years was not reduced in the Alpha-Tocopherol, Beta-Carotene study.

Table: Cross-country correlations between coronary heart disease mortality in men and the supply of vitamin E homologues across 24 European countries

Homologue	Correlation coefficient, r
Total vitamin E	-0. 386
d-a-tocopherol	-0. 753
d-b-tocopherol	-0. 345
d-g-tocopherol	-0. 001
d-d-tocopherol	0. 098
d-a-tocotrienol	-0. 072
d-b-tocotrienol	-0. 329
d-g-tocotrienol	-0. 210

The correlation with d-a-tocopherol is highly significant ($P < 0.001$) whereas all other correlations do not achieve statistical significance.

Source: Based on Reference 30.

It has also been suggested vitamin E supplementation (200-400 mg/day) may be appropriate therapeutically to moderate some aspects of degenerative diseases such as Parkinson's disease, reduce the severity of neurologic disorders such as tardive dyskinesia, prevent periventricular haemorrhage in pre-term babies, reduce tissue injury arising from ischaemia and reperfusion during surgery, delay cataract development, and improve mobility in arthritis sufferers. However, very high doses may also induce adverse pro-oxidant effects, and the long-term advantages of such treatments have not been proven.

Delineation of dietary sources and possible limitations to its availability worldwide.

Because vitamin E is naturally present in plant-based diets and animal products and is often added by manufacturers to vegetable oils and processed foods, intakes are probably adequate to avoid overt deficiency in most situations. Exceptions may be during ecologic disasters and cultural conflicts resulting in food deprivation and famine.

Analysis of the Food and Agriculture Organization of the United Nations country food balance sheets indicates that about half the a-tocopherol in a typical northern European diet such as in the United Kingdom is derived from vegetable oils. Animal fats, vegetables, and meats each contribute about 10 percent to the total per capita supply and fruit, nuts, cereals, and dairy products each contribute about 4 percent. Less than 2 percent is each obtained from eggs, fish and pulses.

There are marked differences in per capita a-tocopherol supply among different countries ranging from approximately 8-10 mg/head/day (e. g., Iceland, Finland, New Sealand, and Japan) to 20-25 mg/head/day. This variation can be ascribed mainly to the type and quantity of dietary oils used in different countries and the proportion of the different homologues in the oils.

For example, sunflower seed oil contains approximately 55 mg a-tocopherol/100 g in contrast to soybean oil that contains only 8 mg/100 ml. Consumption of these oils varies markedly among countries. Soybean, a rich source of the less biologically active g form, is most commonly used in northern European countries whereas sunflower seed oils, which mainly contain the a form, are generally used in southern Europe.

Table: Vitamin E content (mg tocopherol/100g) in vegetable oils

Oil	a-tocopherol	g-tocopherol	d-tocopherol	a-tocotrienol
Coconut	0. 5	0	0. 6	0. 5
Maise (Corn)	11. 2	60. 2	1. 8	0
Palm	25. 6	31. 6	7. 0	14. 3
Olive	5. 1	Trace	0	0
Peanut	13. 0	21. 4	2. 1	0
Soybean	10. 1	59. 3	26. 4	0
Wheatgerm	133. 0	26. 0	27. 1	2. 6
Sunflower	48. 7	5. 1	0. 8	0

EVIDENCE FOR DETERMINING RECOMMENDED NUTRIENT

It was decided that there was insufficient evidence to enable a recommended nutrient intake (RNI) to be based on the additional health benefits obtainable from nutrient intakes above those usually found in the diet. Even for vitamin E with its important biologic antioxidant properties, there was no consistent evidence for protection against chronic disease from dietary supplements. Nevertheless, the main function of vitamin E appears to be that of preventing oxidation of PUFAs, and this has been used by those bodies proposing RNIs for vitamin E because there is considerable evidence in different animal species that low vitamin E and PUFAs excess gives rise to a wide variety of clinical signs.

There is very little clinical evidence of deficiency disease in humans except in certain inherited conditions where the metabolism of vitamin E is disturbed. Even biochemical evidence of poor vitamin E status in both adults and children is minimal. Meta-analysis of data collected within European countries indicates that optimum intakes may be implied when plasma concentrations of vitamin E exceed 25-30 mmol/L of lipid-standardised a-tocopherol. However, this approach should be treated with caution, as plasma vitamin E concentrations do not necessarily reflect intakes or tissue reserves because only 1 percent of the body tocopherol may be in the blood and the amount in the circulation is strongly influenced by circulating lipid.

Nevertheless, the lipid-standardised vitamin E concentration (e. g., tocopherol-cholesterol ratio) greater that 2. 25 is believed to represent satisfactory vitamin E status. The erythrocytes of subjects with values below this concentration of vitamin E may show evidence of an increasing tendency to haemolyse when exposed to oxidizing agents and thus such values should be taken as an indication of biochemical deficiency. However, the development of clinical evidence of vitamin E deficiency (e. g., muscle damage or neurologic lesions) can take several years of exposure to extremely low vitamin E levels.

The main factor used to assess the adequacy of vitamin E intakes by the US and UK advisory bodies was the dietary intake of PUFAs. PUFAs are very susceptible to oxidation, and their increased intake without a concomitant increase in vitamin E can lead to a reduction in plasma vitamin E concentrations and to elevations in some indexes of oxidative damage in human volunteers. Generally, however, diets high in PUFAs are also high in vitamin E, and to set a dietary recommendation based on extremes of PUFA intake would deviate considerably from median intakes of vitamin E in most Western populations. Hence 'safe' allowances for the United Kingdom (men 10 and women 7 mg/day) and 'arbitrary' allowances for the United States (men 10 and women 8 mg/

day) for vitamin E intakes approximate the median intakes in those countries.

It is worth noting that there were only 11 (0. 7 percent) subjects out of 1629 adults in the 1986-1987 British Nutrition Survey who had a-tocopherol - cholesterol ratios <2. 25. Furthermore, although the high intake of soybean oil with its high content of g-tocopherol substitutes for the intake of a-tocopherol in the British diet, a comparison of a-tocopherol-cholesterol ratios found almost identical results in two groups of randomly selected, middle-aged adults in Belfast (Northern Ireland) and Toulouse, two countries with very different intakes of a-tocopherol and cardiovascular risk.

It is suggested that when the main PUFA in the diet is linoleic acid, a d-a-tocopherol-PUFA ratio of 0. 4 (expressed as mg tocopherol per g PUFA) is adequate for adult humans, and the ratio has been recommended in the United Kingdom for infant formulas. Use of this ratio to calculate the vitamin E requirements of men and women with energy intakes of 2550 and 1940 kcal/day containing PUFA at 6 percent of the energy intake produced values of 7 and 5 mg/day of a-TEs, respectively.

At present, data are not sufficient to formulate recommendations for vitamin E intake for different age groups except for infancy. There is some indication that new-born infants, particularly if born prematurely, are vulnerable to oxidative stress because of low body stores of vitamin E, impaired absorption, and reduced transport capacity resulting from low concentrations at birth of circulating low-density lipoproteins. However, term infants almost achieve adult plasma vitamin E concentrations in the first week and although the concentration of vitamin E in early human milk can be variable, after 12 days it remains fairly constant at 0. 32 mg TE/100 ml milk (51). Thus a human-milk-fed infant consuming 850 ml would have an intake of 2. 7 mg. It seems reasonable that formula milk should not contain less than 0. 3 mg TE/100 ml of reconstituted feed and not less than 0. 4 mg TE/g PUFA.

No specific recommendations concerning the vitamin E requirements in pregnancy and lactation have been made by other advisory bodies mainly because there is no evidence of vitamin E requirements different from those of other adults and presumably also as the increased energy intake would compensate for the increased needs for infant growth and milk synthesis.

Vitamin E appears to have very low toxicity, and amounts of 100-200 mg of the synthetic all-rac-a-tocopherol are consumed widely as supplements. Evidence of pro-oxidant damage has been associated with the feeding of supplements but usually only at very high doses (e. g., >1000 mg/day).

Similarly, more investigation is required of the growing evidence that inadequate vitamin E status may increase susceptibility to infection particularly by allowing the genomes of certain relatively benign viruses to convert to more virulent strains.

There is an important need to define optimum vitamin E intakes. Intervention trials with morbidity and mortality endpoints may take years to complete. One approach to circumvent this delay may be to assess the effects of different intakes of vitamin E on biomarkers of oxidative damage to lipids, proteins, and DNA because their occurrence in vivo is implicated in many diseases, including vascular disease and certain cancers.

VITAMIN K

V itamin K is an essential fat-soluble micronutrient which is needed for a unique post-translational chemical modification in a small group of proteins with calcium-binding properties, collectively known as vitamin K - dependent proteins or Gla-proteins. Thus far, the only unequivocal role of vitamin K in health is in the maintenance of normal coagulation. The vitamin K - dependent coagulation proteins are synthesised in the liver and comprise factors II, VII, IX, and X, which have a haemostatic role (i. e., they are procoagulants that arrest and prevent bleeding), and proteins C and S, which have an anticoagulant role (i. e., they inhibit the clotting process).

Despite this duality of function, the overriding effect of nutritional vitamin K deficiency is to tip the balance in coagulation towards a bleeding tendency caused by the relative inactivity of the procoagulant proteins. Vitamin K - dependent proteins synthesised by other tissues include the bone protein osteocalcin and matrix Gla protein; their functions remain to be clarified.

Biological Role of Vitamin K

Vitamin K is the family name for a series of fat-soluble compounds, which have a common 2-methyl-1, 4-naphthoquinone nucleus but differ in the structures of a side chain at the 3-position. They are synthesised by plants and bacteria. In plants the only important molecular form is phylloquinone (vitamin K_1), which has a phytyl side chain. Bacteria synthesise a family of compounds called menaquinones (vitamin K_2), which have side chains based on repeating unsaturated 5-carbon (prenyl) units. These are designated menaquinone-n (MK-n) according to the number (n) of prenyl units. Some bacteria also synthesise menaquinones in which one or more of the double bonds is saturated. The compound 2-methyl-1,4-naphthoquinone (common name menadione) may be regarded as a provitamin because vertebrates can convert it to MK-4 by adding a 4-prenyl side chain at the 3-position.

The biologic role of vitamin K is to act as a cofactor for a specific carboxylation reaction that transforms selective glutamate (Glu) residues to gg-carboxyglutamate (Gla) residues. The reaction is catalysed by a microsomal enzyme, gg-glutamyl, or vitamin K - dependent carboxylase, which in turn is linked to a cyclic salvage pathway known as the vitamin K epoxide cycle.

Scheme shows the cyclic metabolism of vitamin K in relation to the conversion of glutamate (Glu) to gg-carboxyglutamate (Gla) residues for the coagulation protein prothrombin. A general term for the glutamate precursors of vitamin K-dependent proteins is proteins induced by vitamin K absence, abbreviated PIVKA. For prothrombin (factor II) the glutamate precursor is known as PIVKA-II.

The active form of vitamin K needed for carboxylation is the reduced form, vitamin K quinol. Known enzyme reactions are numbered 1, 2, and 3. The carboxylation reaction is driven by a vitamin K-dependent carboxylase activity (reaction 1) which simultaneously converts vitamin K quinol to vitamin K 2,3-epoxide. Vitamin K 2,3-epoxide is reduced back to the quinone and then to the quinol by vitamin K epoxide reductase (reaction 2). The reductase activity denoted 2 is dithiol dependent and is inhibited by coumarin anticoagulants such as warfarin. Dietary vitamin K may enter the cycle via an NAD(P)H-dependent vitamin K reductase activity (reaction 3), which is not inhibited by warfarin.

The four vitamin K-dependent procoagulants (factor II or prothrombin, and factors VII, IX, and X) are serine proteases that are synthesised in the liver and then secreted into the circulation as inactive forms (zymogens). Their biologic activity depends on their normal complement of Gla residues, which are efficient chelators of calcium ions. In the presence of Gla and calcium ions these proteins bind to the surface membrane phospholipids of platelets and endothelial cells where, together with other cofactors, they form membrane-bound enzyme complexes.

When coagulation is initiated, the zymogens of the four vitamin K-dependent clotting factors are cleaved to yield the active protease clotting factors. Two other vitamin K-dependent proteins called protein C and protein S play a regulatory role in the inhibition of coagulation. The function of protein C is to degrade phospholipid-bound activated factors V and VIII in the presence of calcium. Protein S acts as a synergistic cofactor to protein C by enhancing the binding of activated protein C to negatively charged phospholipids. Yet another vitamin K-dependent plasma protein (protein Z) is suspected to have a haemostatic role but its function is currently unknown.

Apart from the coagulation proteins, several other vitamin K-dependent proteins have been isolated from bone, cartilage, kidney, lungs, and other tissues. Only two, osteocalcin and matrix Gla protein (MGP), have been well characterised. Both are found in bone but MGP also occurs in cartilage, blood vessel walls, and other soft tissues. There is evidence that protein S is synthesised by several tissues including the vessel wall and bone and may have other functions besides its well-established role as a coagulation inhibitor.

It also seems likely that one function of MGP is to inhibit mineralisation. Thus far, no clear biologic role for osteocalcin has been established despite its being the major non-collagenous bone protein synthesised by osteoblasts. This failure to establish a biologic function for osteocalcin has hampered studies of the possible detrimental effects of vitamin K deficiency on bone health. Evidence of a possible association of a suboptimal vitamin K status with increased fracture risk remains to be confirmed.

Overview of Metabolism

Absorption and Transport

Dietary vitamin K, mainly as phylloquinone, is absorbed chemically unchanged from the proximal intestine after solubilisation into mixed micelles composed of bile salts and the products of pancreatic lipolysis. In healthy adults the efficiency of absorption of phylloquinone in its free form is about 80 percent. Within the intestinal mucosa the vitamin is incorporated into chylomicrons, is secreted into the lymph, and enters the blood via the lacteals.

Once in the circulation, phylloquinone is rapidly cleared at a rate consistent with its continuing association with chylomicrons and the chylomicron remnants which are produced by lipoprotein lipase hydrolysis at the surface of capillary endothelial cells. After an overnight fast, more than half of the circulating phylloquinone is still associated with triglyceride-rich lipoproteins, with the remainder being equally distributed between low-density and high-density lipoproteins.

Phylloquinone is the major circulating form of vitamin K but MK-7 is present in plasma at lower concentrations and has a lipoprotein distribution similar to phylloquinone. Although phylloquinone in blood must have been derived exclusively from the diet, it is not known whether circulating menaquinones such as MK-7 are derived from the diet, intestinal flora, or a combination of these sources.

Tissue Stores and Distribution

Until the 1970s, the liver was the only known site of synthesis of vitamin K-dependent proteins and hence was presumed to be the only significant storage site for the vitamin. However, the discovery of vitamin K-dependent processes and proteins in a number of extra-hepatic tissues suggests that this may not be the case.

Human liver stores normally comprise about 90 percent menaquinones and 10 percent phylloquinone. There is evidence that the phylloquinone liver stores are very labile; under conditions of severe dietary depletion, liver concentrations were reduced to about 25 percent of initial levels after only 3 days. This high turnover of hepatic reserves of phylloquinone is in accord with the high losses of this vitamer through excretion. Knowledge of hepatic stores of phylloquinone in different population groups is limited. Adult hepatic stores in a UK study were about 11 pmol/g whereas in a study from Japan they were about twofold higher. Such reserves are about 20 000-40 000-fold lower than those for retinol for relative daily intakes of phylloquinone that are only about 10-fold lower than those of vitamin A.

The relationship between hepatic and total-body stores of vitamin K is not known. Other sites of storage may be adipose tissue and bone; both are known to be sites where vitamin K-bearing chylomicrons and chylomicron remnants may be taken up. It has been reported that the predominant vitamer in human cortical and trabecular bone is phylloquinone; unlike the situation in liver, no menaquinones higher than MK-8 were detected.

In contrast to the hepatic preponderance of long-chain menaquinones, the major circulating form of vitamin K is invariably phylloquinone. The menaquinones MK-7 and possibly MK-8 are also present but the common hepatic forms MKs 9-13 are not detectable in blood plasma. This might be a consequence of a different route of absorption (e. g., the possibility of a portal route for long-chain MKs versus the established lymphatic route for phylloquinone) but might suggest that once in the liver, the lipophilic long-chain menaquinones are not easily mobilised.

Bio-activity

Very little information exists on the relative effectiveness of different hepatic forms of K vitamins for the coagulation function of vitamin K in humans. This information is important because of the preponderance of long-chain menaquinones in human liver. Early bioassay data from rats suggested that long-chain menaquinones (MKs-7, 9, and 10) were more efficient than phylloquinone in reversing vitamin K deficiency when single doses were give parenterally and that their sustained response may be due to their slower hepatic turnover. A longer duration of the biologic response of MK-9 compared with phylloquinone in vitamin K-deficient rats was also observed. When given orally, the dietary requirement of MK-9 for the maintenance of prothrombin synthesis in rats is higher than that for phylloquinone. They also reported that the initial hepatic turnover of MK-9 was two- to three-fold slower than that of phylloquinone.

Suttie emphasised that the existence of a large pool of menaquinones in human liver does not necessarily mean that menaquinones make a proportionately greater contribution to the maintenance of vitamin K sufficiency. In humans the development of sub-clinical signs of vitamin K deficiency detected in dietary phylloquinone restriction studies argues against this, especially when placed alongside the lack of change of hepatic menaquinone stores. One explanation is that much of the hepatic menaquinones is not biologically available to the microsomal g-glutamyl carboxylase because of a different sub-cellular location, especially location in the mitochondria and possibly other non-microsomal sites.

Excretion

Vitamin K is extensively metabolised in the liver and excreted in the urine and bile. In tracer experiments it was found that about 20 percent of an injected dose of phylloquinone was recovered in the urine whereas about 40-50 percent was excreted in the faeces via the bile; the proportion excreted was the same regardless of whether the injected dose was 1 mg or 45 μg. It seems likely, therefore, that about 60-70 percent of the amounts of phylloquinone absorbed from each meal will ultimately be lost to the body by excretion. These results suggest that the body stores of phylloquinone are being constantly replenished.

Two major human excretion products have been identified: carboxylic acids with 5 and 7-carbon sidechains that are excreted in the urine as glucuronide conjugates. The biliary metabolites have not been clearly identified but are initially excreted as water-soluble conjugates and become lipid soluble during their passage through the gut, probably through deconjugation by the gut flora. There is no evidence for body stores of vitamin K being conserved by an enterohepatic circulation. Vitamin K itself is too lipophilic to be excreted in the bile and the sidechain-shortened carboxylic acid metabolites are not biologically active.

Populations at Risk

Vitamin K Deficiency Bleeding in Infants

In infants up to around age 6 months, vitamin K deficiency, although rare, represents a significant public health problem throughout the world. The deficiency syndrome is traditionally known as haemorrhagic disease of the newborn or more recently, to give a better definition of the cause, vitamin K deficiency bleeding (VKDB).

The time of onset of VKDB is more unpredictable than previously supposed and it is now useful to recognise three syndromes: early, classic, and late. Until the 1960s, VKDB was considered to be solely a problem of the first week of life. Then, in 1966, came the first reports from Thailand of a new vitamin K deficiency syndrome that typically presented between 1 and 2 months of life and is now termed late VKDB. In later nineties Bhanchet and colleagues, who had first described this syndrome, summarised their studies of 93 affected Thai infants, establishing the idiopathic history, preponderance of breast-fed infants, and high incidence of intracranial bleeding.

Without vitamin K prophylaxis, the incidence of late VKDB (per 100,000 births), based on acceptable surveillance data, has been estimated to be 4. 4 in the United Kingdom, 7. 2 in Germany, and as high as 72 in Thailand. Of real concern is that late VKDB, unlike the classic form, has a high incidence of death or severe and permanent brain damage resulting from intracranial haemorrhage.

Epidemiologic studies worldwide have identified two major risk factors for both classic and late VKDB: exclusive human milk feeding and the failure to give any vitamin K prophylaxis. The increased risk for infants fed human milk compared with formula milk is probably related to the relatively low concentrations of vitamin K (phylloquinone) in breast milk compared with formula milks.

For classic VKDB, studies using the detection of under-carboxylated prothrombin or proteins induced by vitamin K absence (PIVKA)-II as a marker of sub-clinical vitamin K deficiency have suggested that it is the low cumulative intake of human milk in the first week of life rather than an abnormally low milk concentration per se that seems to be of greater relevance. Thus, classic VKDB may be related, at least in part, to a failure to establish early breast-feeding.

Vitamin K Prophylaxis in Infants

Because bleeding can occur spontaneously and because no screening test is available, it is now common paediatric practice to protect all infants by giving vitamin K supplements in the immediate perinatal period. Vitamin K prophylaxis has had a chequered history but in recent years has become a high-profile issue of public health in many countries throughout the world. The reasons for this are twofold. First there is now a convincing body of evidence showing that without vitamin K prophylaxis, infants have a small but real risk of dying from or being permanently brain damaged by vitamin K deficiency in the first 6 months of life.

The other, much less certain evidence stems from a reported epidemiologic association between vitamin K given intramuscularly (but not orally) and the later development of childhood cancer. The debate, both scientific and public, which followed this and other publications has led to an increase in the use of multiple oral supplements instead of the traditional single intramuscular injection (usually of 1 mg of phylloquinone) given at birth. Although most of the subsequent epidemiologic studies have not confirmed any cancer link with vitamin K, the issue is still not resolved.

Vitamin K in Adults

In adults, primary vitamin K-deficient states that manifest as bleeding are almost unknown except when the absorption of the vitamin is impaired as a result of an underlying pathology.

Dietary Sources

High-performance liquid chromatography can be used to accurately determine the major dietary form of vitamin K (phylloquinone) in foods, and food tables are being compiled for Western diets. Phylloquinone is distributed ubiquitously throughout the diet, and the range of concentrations in different food categories is very wide. In general, the relative values in vegetables confirm the known association of phylloquinone with photosynthetic tissues, with the highest values (normally in the range 400-700 μg/100 g) being found in green leafy vegetables. The next best sources are certain vegetable oils (e. g., soybean, rapeseed, and olive oils) which contain 50-200 μg/100 g. Some vegetable oils, such as peanut, corn, sunflower and safflower oils, have a much lower phylloquinone content (1-10 μg/100 g). The great differences between vegetable oils obviously presents problems for calculating the phylloquinone contents of oil-containing foods when the type of oil (or its storage condition) is not known.

Menaquinones seem to have a more restricted distribution in the diet than does phylloquinone. In the Western diet nutritionally significant amounts of long-chain menaquinones have been found in animal livers and fermented foods such as cheeses. Yeasts do not synthesise menaquinones and menaquinone-rich foods are those with a bacterial fermentation stage. The Japanese food natto (fermented soybeans) has a menaquinone content even higher than that of phylloquinone in green leafy vegetables.

The relative dietary importance of MK-4 is more difficult to evaluate because concentrations in foods may well depend on geographic differences in the use of menadione in animal husbandry, menadione from which MK-4 may be synthesised in animal tissues. Another imponderable factor is the evidence that animal tissues and dairy produce may contain some MK-4 as a product of tissue synthesis from phylloquinone itself.

Knowledge of the vitamin K content of human milk has been the subject of methodologic controversies with a 10-fold variation in reported values of phylloquinone concentrations of mature human milk. Where milk sampling and analytical techniques have met certain criteria for their validity, the phylloquinone content of mature milk have generally ranged between 1 and 4 μg/l, with average concentrations near the lower end of this range. However, there is considerable intra- and inter-subject variation, and levels higher are in colostral milk than in mature milk. Menaquinone concentrations in human milk have not been accurately determined but appear to be much lower than those of phylloquinone. Phylloquinone concentrations in infant formula milk range from 3 to 16 μg/l in unsupplemented formulas and up to 100 μg/l in fortified formulas. Nowadays most formulas are fortified; typical phylloquinone concentrations are about 50 mg/l.

BIO-AVAILABILITY OF VITAMIN K FROM FOODS

Very little is known about the bio-availability of the K vitamins from different foods. It has been estimated that the efficiency of absorption of phylloquinone from boiled spinach (eaten with butter) is no greater than 10 percent compared with an estimated 80 percent when phylloquinone is given in its free form. This poor absorption of phylloquinone from green leafy vegetables may be explained by its location in chloroplasts and tight association with the thylakoid membrane, where this naphthoquinone plays a role in photosynthesis. In comparison, the bio-availability of MK-4 from butter artificially enriched with this vitamer was more than twofold higher than that of phylloquinone from spinach.

The poor extraction of phylloquinone from leafy vegetables, which as a category represents the single greatest food source of phylloquinone, may place a different perspective on the relative importance of other foods with lower concentrations of phylloquinone but in which the vitamin is not tightly bound and its bio-availability is likely to be greater. Even before bio-availability was taken into account, fats and oils that are contained in mixed dishes were found to make an important contribution to the phylloquinone content of the US diet and in a UK study contributed 30 percent of the total dietary intake.

Importance of Intestinal Bacterial Synthesis as a Source of Vitamin K

Intestinal microflora synthesise large amounts of menaquinones, which are potentially available as a source of vitamin K. Quantitative measurements at different sites of the human intestine have demonstrated that most of these menaquinones are present in the distal colon. Major forms produced are MK-10 and MK-11 by Bacteroides, MK-8 by Enterobacter, MK-7 by Veillonella, and MK-6 by Eubacterium lentum. It is noteworthy that menaquinones with very long chains (MKs 10-13) are known to be synthesised by members of the anaerobic genus Bacteroides and are major inhabitants of the intestinal tract but have not been detected in significant amounts in foods. The widespread presence of MKs 10-13 in human livers at high concentrations therefore suggests that these forms, at least, originate from intestinal synthesis.

It is commonly held that animals and humans obtain a significant fraction of their vitamin K requirement from direct absorption of menaquinones produced by microfloral synthesis, but hard experimental evidence documenting the site and extent of any absorption is singularly lacking. The most promising site of absorption is the terminal ileum, where there are some menaquinone-producing bacteria as well as bile salts. The evidence overall suggests that the bio-availability of bacterial menaquinones is poor because they are mostly tightly bound to the bacterial cytoplasmic membrane and the largest pool is present in the colon, which lacks bile salts for their solubilisation.

Evidence on Which Recommendations can be Based

Assessment of Vitamin K Status

Conventional coagulation assays are useful for detecting overt vitamin K-deficient states which are associated with a risk of bleeding. However, they offer only a relatively insensitive insight into vitamin K nutritional status and the detection of sub-clinical vitamin K-deficient states. A more sensitive measure of vitamin K sufficiency can be obtained from tests that detect under-carboxylated species of vitamin K-dependent proteins. In states of vitamin K deficiency, under-carboxylated species of the vitamin K-dependent coagulation proteins are released from the liver into the blood; their levels increase with the degree of severity of vitamin K deficiency. These under-carboxylated forms (PIVKA) are unable to participate in the normal coagulation cascade because they are unable to bind calcium.

The measurement of under-carboxylated prothrombin (PIVKA-II) is the most useful and sensitive homeostatic marker of sub-clinical vitamin K deficiency. Importantly, PIVKA-II is detectable in plasma before any changes occur in conventional coagulation tests. Several types of assay for PIVKA-II have been developed which vary in their sensitivity. In the same way that vitamin K deficiency causes PIVKA-II to be released into the circulation from the liver, a deficit of vitamin K in bone will cause the osteoblasts to secrete under-carboxylated species of osteocalcin (ucOC) into the bloodstream. It has been proposed that the concentration of circulating ucOC reflects the sufficiency of vitamin K for the carboxylation of this Gla protein in bone tissue. Most assays for ucOC have been indirect because they rely on the differential absorption of carboxylated and under-carboxylated forms to hydroxyapatite and are difficult to interpret.

Other criteria of vitamin K sufficiency that have been used are plasma measurements of phylloquinone and the measurement of urinary Gla. It is expected and found that the excretion of urinary Gla is decreased in vitamin K deficiency.

Dietary Intakes in Infants and their Adequacy

The average intake of phylloquinone in infants fed human milk during the first 6 months of life has been reported to be less than 1 μg/day; this is approximately 100-fold lower than the intake in infants fed a typical supplemented formula. This big disparity between intakes is reflected in plasma levels.

Using the detection of PIVKA-II as a marker of sub-clinical deficiency, a study from Germany concluded that a minimum daily intake of about 100 ml of colostral milk (that supplies about 0. 2-0. 3 μg of phylloquinone) is sufficient for normal haemostasis in a baby of about 3 kg during the first week of life. Similar conclusions were reached in a Japanese study which showed a linear correlation between the prevalence of PIVKA-II and the volume of breast milk ingested over 3 days; 95 percent of infants with detectable PIVKA-II had average daily intakes of less than about 120 ml, but the marker was not detectable when intakes reached 170 ml/day.

FACTORS OF RELEVANCE TO CLASSICAL VITAMIN K DEFICIENCY BLEEDING

The liver stores of vitamin K in the newborn infant differ both qualitatively and quantitatively from that in adults. First, phylloquinone levels at birth are about one-fifth those in adults and second, bacterial menaquinones are undetectable. The resistance to placental transport of vitamin K to the human foetus is well-established. Thereafter, the limited available data suggest that hepatic stores of menaquinones build up gradually, becoming detectable at around the second week of life but only reaching adult concentrations after 1 month of age. A gradual increase in liver stores of menaquinones may reflect the gradual colonisation of the gut by enteric microflora.

A practical problem in assessing the functional status of vitamin K in the neonatal period is that there are both gestational and postnatal increases in the four vitamin K-dependent procoagulant factors which are unrelated to vitamin K status (50). This means that unless the deficiency state is quite severe, it is very difficult to interpret clotting factor activities as a measure of vitamin K sufficiency. The best diagnostic tool of the adequacy of vitamin K stores for neonates is by the detection of PIVKA-II by immunoassays.

The use of this marker has clearly shown that there is a temporary dip in the vitamin K status of infants exclusively fed human milk in the first few days after birth. The fact that the degree of this dip is associated with human-milk intakes and is less evident or abolished in infants given artificial feeds or prophylactic vitamin K at birth shows that the detection of PIVKA-II reflected a dietary lack of vitamin K.

Factors of Relevance to Late Vitamin K Deficiency bleeding

The natural tendency for human-milk-fed infants to develop a sub-clinical vitamin K deficiency in the first 2-3 days of life is self-limiting. Comparisons between untreated human-milk-fed infants with others who had received vitamin K or supplementary feeds clearly suggest that this improvement in vitamin K-dependent clotting activity is due to an improved vitamin K status. After the first week, vitamin K-dependent clotting factor increases are more gradual, and it is not possible from clotting factor assays to differentiate between the natural post-natal increase in the synthesis of the core proteins from an improved vitamin K status leading to greater functional activity.

Use of the most sensitive assays for PIVKA-II shows that there is still evidence of suboptimal vitamin K status in infants solely fed human milk between the ages of 1 and 2 months. Deficiency signs are less common in infants who have received adequate vitamin K supplementation or who have been formula fed.

DIETARY INTAKES IN OLDER INFANTS, CHILDREN, AND ADULTS AND THEIR ADEQUACY

The only comprehensive national survey of phylloquinone intakes across all age groups (except infants aged 0-6 months) is that of the US Food and Drug Administration Total Diet Study, which was based on the 1987-88 Nationwide Food Consumption Survey. For infants and children from the age of 6 months to 16 years, average phylloquinone intakes were above the current US recommended dietary allowance (RDA) values for their respective age groups, more so for children up to 10 years than from 10 to 16 years. There have been no studies of intakes in children in relation to functional markers of vitamin K sufficiency in children.

Intakes for adults in The Total Diet Study were also close to or slightly higher than the current US RDA values of 80 μg for men and 65 μg for women, although intakes were slightly lower than the RDA in the 25-30 years age group (54). There is some evidence from an evaluation of all the US studies that older adults have higher dietary intakes of phylloquinone than do younger adults.

The US results are very comparable with a detailed, seasonality study in the United Kingdom in which mean intakes in men and women (aged 22-54 years) were 72 and 64 μg/day, respectively; no significant sex or seasonal variations were found. Another UK study suggested that intakes were lower in manual workers and in smokers, reflecting their lower intakes of green vegetables and high-quality vegetable oil.

Several dietary restriction and repletion studies have attempted to assess the adequacy of vitamin K intakes in adults (55, 58). It is clear from these studies that volunteers consuming less than 10 μg/day of phylloquinone do not show any changes in conventional coagulation tests even after several weeks unless other measures to reduce the efficiency of absorption are introduced. However, a diet containing only 2-5 μg/day of phylloquinone fed for 2 weeks did result in an increase of PIVKA-II and a 70 percent decrease in plasma phylloquinone (59). Similar evidence of a sub-clinical vitamin K deficiency together with an increased urinary excretion of Gla was found when dietary intakes of phylloquinone were reduced from about 80 to about 40 μg/day for 21 days. A repletion phase in this study was consistent with a human dietary vitamin K requirement (for its coagulation role) of about 1 μg/kg body weight/day.

The most detailed and controlled dietary restriction and repletion study in healthy human subjects. In this study 32 healthy subjects in two age groups (20-40 and 60-80 years) were fed a mixed diet containing about 80 μg/day of phylloquinone, which is the RDA for adult males in the United States (54). After 4 days on this baseline diet there was a 13-day depletion period during which the subjects were fed a diet containing about 10 μg/day. After this depletion phase the subjects entered a 16-day repletion period during which, over 4-day intervals, they were sequentially repleted with 5, 15, 25, and 45 μg phylloquinone. The depletion protocol had no effect on conventional coagulation and specific factor assays but did induce a significant increase in PIVKA-II in both age groups. The most dramatic change was in plasma levels of phylloquinone, which fell to about 15 percent of the values on day 1.

The drop in plasma phylloquinone also suggested that the average dietary intake of these particular individuals before they entered the study had been greater than the baseline diet of 80 μg/day. The repletion protocol failed to bring the plasma phylloquinone levels of the young subjects back above the lower limit of the normal range (previously established in healthy, free-living adults) whereas the plasma levels in the elderly group only rose slightly above this lower limit in the last 4 days. Another indication of a reduced vitamin K status in the young group was the fall in urinary output of Gla (90 percent of baseline) that was not seen in the elderly group; this suggested that younger subjects are more susceptible to the effects of an acute deficiency than are older subjects.

One important dietary intervention study measured the carboxylation status of the bone vitamin K-dependent protein, osteocalcin, in response to altered dietary intakes of phylloquinone. This was a crossover study, which evaluated the effect in young adults of increasing the dietary intake of phylloquinone to 420 μg/day for 5 days from a baseline intake of 100 μg/day. Although total concentrations of osteocalcin were not affected by either of the dietary treatments, ucOC fell dramatically in response to the 420 μg diet and by the end of the 5-day supplementation period was 41 percent lower than the baseline value. After the return to the mixed diet, the ucOC percent rose significantly but after 5 days had not returned to pre-supplementation values. This study suggested that the carboxylation of osteocalcin in bone may require higher dietary intakes of vitamin K than those needed to sustain its haemostatic function.

Recommendations for Vitamin K Intakes

Table: Recommended nutrients intakes for vitamin K

Age group	Recommended Nutrient Intake† μg /day
Infants and children	
0-6 months	5*
7-12 months	10
1-3 years	15
4-6 years	20
7-9 years	25
Adolescents, 10-18 years	
Females	35-55
Males	35-55
Adults	
Females, 19-65 years	55
65+ years	55
Males, 19-65 years	65
65+ years	65
Pregnancy	55
Lactation	55

To prevent bleeding due to vitamin K deficiency, the panel recommends that all breast babies should receive vitamin K supplementation at birth according to nationally approved guidelines. Vitamin K formulations and prophylactic regimes differ from country to country. Guidelines range from a single intramuscular injection (usually 1 mg of phylloquinone) given at birth to multiple oral doses given over the first few weeks.

Infants 0-6 Months

Consideration of the requirements of vitamin K for infants up to age 6 months is complicated by the need to prevent a rare but potentially devastating bleeding disorder, which is caused by vitamin K deficiency. To protect the few affected infants, most developed and some developing countries have instituted a blanket prophylactic policy to protect infants at risk. The numbers of infants at risk without such a programme has a geographic component, being more prevalent in the Far East, and a dietary component with solely human-milk-fed babies having the highest risk. Of the etiologic factors, some of which may still be unrecognised, one factor in some infants is mild cholestasis.

The problem of overcoming a variable and, in some infants, inefficient absorption is the likely reason that oral prophylactic regimens, even with two or three pharmacologic doses (1 mg phylloquinone), have occasionally failed to prevent VKDB. This makes it difficult to design an effective oral prophylaxis regimen that is comparable in efficacy with the previous "gold standard" of 1 mg phylloquinone given by intramuscular injection at birth. In several countries intramuscular prophylaxis fell out of favour after the epidemiologic report and subsequent controversy that this route may be linked to childhood cancer.

Infants who have been entirely fed with supplemented formulas are well protected against VKDB and on intakes of around 50 µg/day have plasma levels that are about 10-fold higher than the adult average of about 1. 0 nmol/l (0. 5 µg/l). Clearly then, an optimal intake would lie below an intake of 50 µg/day. The effectiveness of giving infants a daily supplement of 25 µg phylloquinone after they had received a single oral dose of 1 µg at birth. This regimen resulted in median plasma levels at ages 4, 8, and 12 weeks of around 2. 2 nmol/l (1. 0 µg/l) when sampled 20-28 hours after the most recent vitamin K dose; this level corresponds to the upper end of the adult fasting range. In 12-week-old infants supplemented with this regime, the median plasma level was about fourfold higher than that in a control group of unsupplemented infants (1. 9 versus 0. 5 nmol/l). Also none of the 50 supplemented infants had detectable PIVKA-II at 12 weeks compared with 15 of 131 infants (11. 5 percent) in the control group. This regime has now been implemented in The Netherlands and surveillance data on late VKDB suggest that it may be as effective as parenteral vitamin K prophylaxis.

The fact that VKDB is epidemiologically associated with breast feeding means that it is not prudent to base requirements solely on normal intakes of human milk and justifies the setting of a higher value that can only be met by some form of supplementation. The current US RDA for infants is 5 μg/day for the first 6 months (the greatest period of risk for VKDB) and 10 μg/day during the second 6 months (54). This is based on the adult RDA of 1 μg/kg body weight/day.

However, if the vitamin K content of human milk is assumed to be about 2 μg/l, exclusively breast-fed infants aged 0-6 months may ingest only 20 percent of their presumed daily requirement of 5 μg (54). Whether a figure of 5 μg/day is itself safe is uncertain. In the United Kingdom the dietary reference value for infants was set at 10 μg/day, which in relation to body weight (2 μg/kg) is about double the estimate for adults. It was set with reference to the upper end of possible human milk concentrations plus a further qualitative addition to allow for the absence of hepatic menaquinones in early life and the presumed reliance on dietary vitamin K alone.

The association of VKDB with breast-feeding does not mean that most infants are at risk of developing VKDB, because this is a rare vitamin K deficiency syndrome. In contrast to measurements of PIVKA-II levels, comparisons of vitamin K-dependent clotting activities have shown no detectable differences between infants fed human milk and those fed artifical formula. The detection of PIVKA-II with normal functional levels of vitamin K-dependent coagulation factors does not imply immediate or even future haemorrhagic risk for a particular individual. The major value of PIVKA-II measurements in infants is to assess the prevalence of suboptimal vitamin K status in population studies. However, because of the potential consequences of VKDB, the paediatric profession of most countries agrees that some form of vitamin K supplementation is necessary even though there are widespread differences in actual practice.

Infants (7-12 months), Children, and Adults

In the past, the requirements for vitamin K only considered its classical function in coagulation; an RDA was given for vitamin K in the United States (54, 58) and a safe and adequate intake level was given in the United Kingdom.

In both countries the adult RDA or adequate intake was set at a value of 1 μg/kg body weight/day. Thus in the United States the RDA for a 79-kg man is listed as 80 μg/day and for a 63-kg women as 65 μg/day.

At the time previous recommendations were set there were few data on dietary intakes of vitamin K (mainly phylloquinone) in different populations. The development of more accurate and wide-ranging food databases is now helping to address this question. The results of several dietary intake studies in the United States and the United Kingdom suggest that the average intakes for adults are very close to the respective recommendations of each country.

In the United States, preliminary intake data also suggest that average intakes of phylloquinone in children and adolescents also exceed the RDA; in 6-month-old infants the intakes exceeded the RDA of 10 μg by nearly eightfold, reflecting the use of supplemented formula foods. Because there is no evidence of even sub-clinical deficiencies of haemostatic function, a daily intake of 1 μg/kg may still be used as the basis for the recommended nutrient intake (RNI). There is no basis as yet for making different recommendations for pregnant and lactating women.

A relevant question is whether the RNI should be raised to take into account recent evidence that the requirements for the optimal carboxylation of vitamin K-dependent proteins in other tissues are greater than those for coagulation. There is certainly evidence that the g-carboxylation of osteocalcin can be improved by intakes somewhere between 100 and 420 μg/day. If an RNI for vitamin K sufficiency is to be defined as that amount necessary for the optimal carboxylation of all vitamin K-dependent proteins, including osteocalcin, then it seems clear that this RNI would lie somewhere above the current intakes of many, if not most, of the population in the United States and the United Kingdom. Because a clearly defined metabolic role and biochemical proof of the necessity for fully g-carboxylated osteocalcin for bone health is currently lacking, it would be unwise to make such a recommendation.

Toxicity

When taken orally, natural K vitamins seem free of toxic side effects. This safety is illustrated by the common clinical administration of phylloquinone at doses of 10-20 mg or greater. Some patients with chronic fat malabsorption regularly ingest doses of this sise without evidence of any harm. However, synthetic preparations of menadione or its salts are best avoided for nutritional purpose, especially for vitamin prophylaxis in the newborn. Besides lacking intrinsic biologic activity, the high reactivity of its unsubstituted 3-position has been associated with neonatal haemolysis and liver damage.

Chapter 5

Role of Minerals

CALCIUM

The rediscovery of earlier information that calcium deficiency led to the development of osteoporosis (not rickets and osteomalacia) in experimental animals resulted in a reexamination of osteoporosis in humans, notably in postmenopausal women. This reexamination yielded evidence in the late 1960s that menopausal bone loss was not due to a decrease in bone formation but rather to an increase in bone resorption (5-8), and this has had a profound effect on our understanding of other forms of osteoporosis.

Although reduced bone formation may aggravate the bone loss process in elderly people and probably plays a major role in corticosteroid osteoporosis and possibly in osteoporosis in men bone resorption is increasingly held responsible for osteoporosis in women and for the bone deficit associated with hip fractures in elderly people of both sexes. Because bone resorption is also the mechanism whereby calcium deficiency destroys bone, it is hardly surprising that the role of calcium in the pathogenesis of osteoporosis has received increasing attention and that recommended calcium intakes have risen steadily in the past 35 years from the nadir which followed the publication of the report from Rome. The process has been accelerated by the growing realisation that insensible losses of calcium (via skin, hair, nails, etc.) need to be taken into account in the calculation of calcium requirement.

As the calcium allowances recommended for developed nations have been rising - and may still not have reached their peak - the gap between them and the actual calcium intakes in developing countries has widened. The concept that calcium requirement may itself vary from culture to culture for dietary, genetic, lifestyle, and geographical reasons is emerging. This report therefore seeks to make it clear that our main recommendations are largely based on data derived from the developed world and are not necessarily applicable to nations with different dietary cultures, different lifestyles, and different environments for which different calculations may be indicated.

Chemistry and Distribution of Calcium

Calcium is a divalent cation with an atomic weight of 40. In the elementary composition of the human body, it ranks fifth after oxygen, carbon, hydrogen, and nitrogen, and it makes up 1. 9 percent of the body by weight. Carcass analyses show that it constitutes 0. 1-0. 2 percent of early foetal fat-free weight, rising to about 2 percent of adult fat-free weight. In absolute terms, this represents a rise from about 24 g (600 mmol) at birth to 1300 g (32. 5 mol) at maturity, requiring an average daily positive calcium balance of 180 mg (4. 5 mmol) during the 20 years of growth.

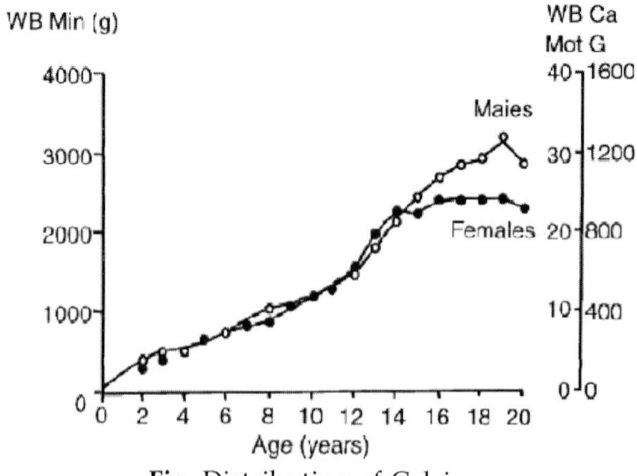

Fig. Distribution of Calcium

Whole-body bone mineral (WB Min) (left axis) and calcium (right axis) as a function of age as determined by total-body dual-energy X-ray absorptiometry

Ninety-nine percent of the body calcium is located in the skeleton. The remaining 1 percent is equally distributed between the teeth and soft tissues, with only 0. 1 percent in the extracellular fluid (ECF). In the skeleton it constitutes 25 percent of the dry weight and 40 percent of the ash weight. The ECF contains ionised calcium at about 4. 8 mg/100 ml (1. 20 mmol/l) maintained by the parathyroid - vitamin D system as well as complexed calcium at about 1. 6 mg/100 ml (0. 4 mmol/l). In the plasma there is an additional protein-bound calcium fraction of 3. 2 mg/100 ml (0. 8 mmol/l). In the cellular compartment the total calcium concentration is comparable with that in the ECF, but the free calcium concentration is lower by several orders of magnitude.

Biological Role of Calcium

Calcium salts provide rigidity to the skeleton and calcium ions play a role in many if not most metabolic processes. In the primitive exoskeleton and in shells, rigidity is generally provided by calcium carbonate, but in the vertebrate skeleton it is provided by a form of calcium phosphate which approximates hydroxyapatite $[Ca_{10}(OH)_2(PO_4)_6]$ and is embedded in collagen fibrils.

Bone mineral serves as the ultimate reservoir for the calcium circulating in the ECF. Calcium enters the ECF from the gut by absorption and from bone by resorption. Calcium leaves the ECF via the gastrointestinal tract, kidneys, and skin and enters into bone via bone formation. In addition, calcium fluxes occur across all cell membranes.

Many neuromuscular and other cellular functions depend on the maintenance of the ionised calcium concentration in the ECF. Calcium fluxes are also important mediators of hormonal effects on target organs through several intracellular signalling pathways, such as the phosphoinositide and cyclic adenosine monophosphate systems. The cytoplasmic calcium concentration is kept down by a series of calcium pumps, which concentrate calcium within the intracellular storage sites or extrude from the cells the calcium which flows in by diffusion.

The physiology of calcium metabolism is primarily directed towards the maintenance of the concentration of ionised calcium in the ECF.

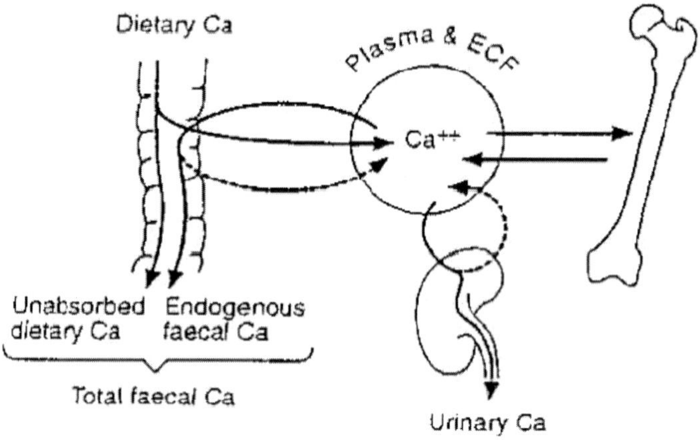

Fig. Major calcium movements in the body

This is protected and maintained by a feedback loop through calcium receptors in the parathyroid glands, which control the secretion of parathyroid hormone. This hormone increases the renal tubular reabsorption of calcium, promotes intestinal calcium absorption by stimulating the renal production of 1,25-dihyroxycolecaliferol [1,25$(OH)_2$D], and, if necessary, resorbs bone. However, the integrity of the system depends critically on vitamin D status; if there is a deficiency of vitamin D, the loss of its calcaemic action leads to a decrease in the ionised calcium and secondary hyperparathyroidism and hypophosphataemia.

This is why experimental vitamin D deficiency results in rickets and osteomalacia whereas calcium deficiency gives rise to osteoporosis.

Determinants of calcium balance

Calcium intake

In a strictly operational sense, calcium balance is determined by the relationship between calcium intake and calcium absorption and excretion. A striking feature of the system is that relatively small changes in calcium absorption and excretion can neutralise a high intake or compensate for a low one.

There is a wide variation in calcium intake among nations, generally following the animal protein intake and depending largely on dairy product consumption. The lowest calcium intakes are in developing countries, particularly in Asia, and the highest are in developed countries, particularly in USA, Canada and Europe.

Table: Protein and calcium intakes in different regions of the world 1987-1989

	Protein (g)			Calcium (mg)		
Region	**Total**	**Animal**	**Vegetable**	**Total**	**Animal**	**Vegetable**
USA and Canada	108. 7	72. 2	36. 5	1031	717	314
Europe	102. 0	59. 6	42. 4	896	684	212
Oceania	98. 3	66. 5	31. 8	836	603	233
Other developed	91. 1	47. 3	43. 8	565	314	251
USSR	106. 2	56. 1	50. 1	751	567	184
All developed	103. 0	60. 1	42. 9	850	617	233
Africa	54. 1	10. 6	43. 5	368	108	260
Latin America	66. 8	28. 6	38. 2	476	305	171
Near East	78. 7	18. 0	60. 7	484	223	261
Far East	58. 2	11. 0	47. 2	305	109	196
Other developing	55. 8	22. 7	33. 1	432	140	292
All developing	59. 9	13. 3	46. 6	344	138	206

Calcium Absorption

Ingested calcium mixes with digestive juice calcium in the proximal small intestine from where it is absorbed by a process, which has an active saturable component and a diffusion component. At low calcium intakes calcium is mainly absorbed by active (transcellular) transport, but at higher intakes an increasing proportion of calcium is absorbed by simple (paracellular) diffusion. The unabsorbed component appears in the faeces together with the unabsorbed component of digestive juice calcium known as endogenous faecal calcium. Thus, the faeces contain unabsorbed dietary calcium and unreabsorbed digestive juice calcium.

True absorbed calcium is the total calcium absorbed from the calcium pool in the intestines and therefore contains both dietary and digestive juice components. Net absorbed calcium is the difference between dietary calcium and faecal calcium and is numerically the same as true absorbed calcium minus endogenous faecal calcium.

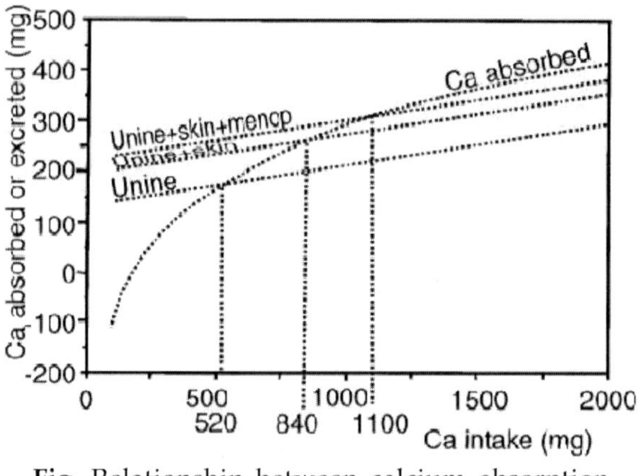

Fig. Relationship between calcium absorption and calcium intake

At sero calcium intake, all the faecal calcium is endogenous and represents the digestive juice calcium which has not been reabsorbed; net absorbed calcium at this intake is therefore negative to the extent of about 200 mg (5 mmol). When the intake reaches about 200 mg (5 mmol), dietary and faecal calcium become equal and net absorbed calcium is sero. As calcium intake increases, net absorbed calcium also increases, steeply at first but then, as the active transport becomes saturated, more slowly until the slope of absorbed on ingested calcium approaches linearity with an ultimate gradient of about 5-10 percent. The relationship between intestinal calcium absorption and calcium intake, derived from 210 balance studies performed in 81 individuals collected from the literature.

The relationships between calcium intake and calcium absorbed and excreted calcium calculated from 210 balance experiments in 81 subjects. Equilibrium is reached at an intake of 520 mg, which rises to 840 mg when skin losses of 60 mg are added and to 1100 mg when menopausal loss is included. The curvilinear relationship between intestinal calcium absorption and calcium intake can be made linear by using the logarithm of calcium intake to yield the equation: Caa = 174 loge Cai -909 ± 71 (SD) mg/day, where Cai represents ingested calcium and Caa net absorbed calcium. The relationship between urinary calcium excretion and calcium intake is given by the equation: Cau = 0. 078 Cai + 137 ± 11. 2 (SD) mg/day, where Cau is urinary calcium and Cai calcium intake.

True absorption is an inverse function of calcium intake, falling from some 70 percent at very low intakes to about 35 percent at high intakes. Percent net absorption is negative at low intakes, becomes positive as intake increases, reaches a peak of about 30 percent at an intake of about 400 mg, and then falls off as the intake increases. The two lines converge as intake rises because the endogenous faecal component (which separates them) becomes proportionately smaller.

Many factors influence the availability of calcium for absorption and the absorptive mechanism itself. The former includes substances, which form insoluble complexes with calcium, such as the phosphate ion. The relatively high calcium-phosphate ratio of 2. 2 in human milk compared with

0. 77 in cow milk may be a factor in the higher absorption of calcium from human milk than cow milk.

Intestinal calcium absorption is mainly controlled by the serum concentration of 1,25$(OH)_2$D. The activity of the 1-a-hydroxylase, which catalyses 1,25$(OH)_2$D production from 25-hydroxycolecalciferol (25OHD) in the kidneys, is negatively related to the plasma calcium and phosphate concentrations and positively to plasma parathyroid hormone. Thus the inverse relationship between calcium intake and fractional absorption described above is enhanced by the inverse relationship between dietary calcium and serum 1,25$(OH)_2$D.

Phytates, present in the husks of many cereals as well as in nuts, seeds, and legumes, can form insoluble calcium phytate salts in the gastrointestinal tract. Excess oxalates can precipitate calcium in the bowel but are not an important factor in most diets.

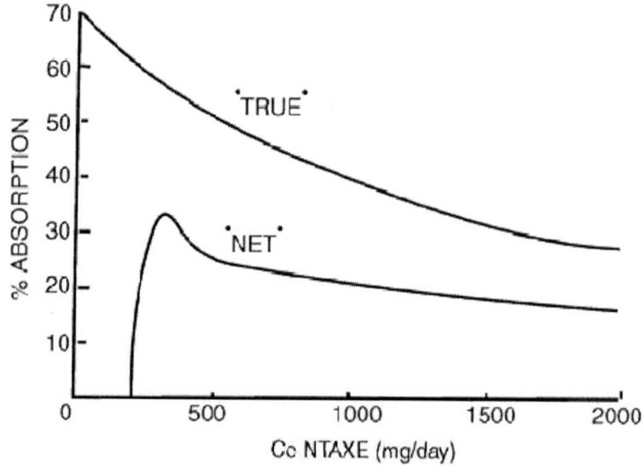

Fig. True and net calcium absorption as percents of calcium intake

Urinary Calcium

Urinary calcium is the fraction of the filtered plasma water calcium, which is not reabsorbed in the renal tubules. At a normal glomerular filtration rate of 120 ml/min and ultrafiltrable calcium of 6. 4 mg/100 ml (1. 60 mmol/l), the filtered load of calcium is about 8 mg/min (0. 20 mmol/min) or 11. 6 g/day (290 mmol/day). Because the usual 24-hour calcium excretion in developed countries is about 160-200 mg (4-5 mmol), it follows that 98-99 percent of the filtered calcium is usually reabsorbed in the renal tubules.

However, calcium excretion is extremely sensitive to changes in filtered load. A decrease in plasma water calcium of only 0. 17 mg/100 ml (0. 043 mmol/l), which is barely detectable, was sufficient to account for a decrease in urinary calcium of 63 mg (1. 51 mmol) when 27 subjects changed from a normal- to a low-calcium diet. This very sensitive renal response to calcium deprivation combines with the inverse relationship between calcium intake and absorption to stabilise the plasma ionised calcium concentration and to preserve the equilibrium between calcium entering and leaving the ECF over a wide range of calcium intakes.

However, there is always a significant obligatory loss of calcium in the urine (as there is in the faeces), even on a low calcium intake, simply because maintenance of the plasma ionised calcium and, therefore, of the filtered load, prevents total elimination of the calcium from the urine. The lower limit for urinary calcium in developed countries is about 140 mg (3. 5 mmol) but depends on protein and salt intakes. From this obligatory minimum, urinary calcium increases on intake with a slope of about 5-10 percent. In the graph derived from 210 balance studies referred to above, the relationship between urinary calcium excretion and calcium intake is represented by the line which intersects the absorbed calcium line at an intake of 520 mg.

Insensible Losses

Urinary and endogenous faecal calcium are not the only forms of excreted calcium; losses through skin, hair, and nails need to be taken into account. These are not easily measured, but a combined balance and isotope procedure has yielded estimates of daily insensible calcium losses in the range of 40-80 mg (1-2 mmol), which are unrelated to calcium intake. The addition of a loss of 60 mg (1. 5 mmol) as a constant to urinary calcium loss raises the dietary calcium at which absorbed and excreted calcium reach equilibrium from 520 to 840 mg (13 to 21 mmol).

Calcium Requirements and Recommended Intakes

Methodology

Although it is well established that calcium deficiency causes osteoporosis in experimental animals, the contribution that calcium deficiency makes to osteoporosis in humans is much more controversial, not least because of the great variation in calcium intakes across the world, which does not appear to be associated with any corresponding variation in the prevalence of osteoporosis. This issue is dealt with at greater length below in the section on nutritional factors; in this section we will simply define what is meant by calcium requirement and how it may be calculated.

The calcium requirement of an adult is generally recognised to be the intake required to maintain calcium balance and therefore skeletal integrity. The mean calcium requirement of adults is therefore the mean intake at which intake and output are equal, which at present can only be determined by balance studies conducted with sufficient care and over a sufficiently long period to ensure reasonable accuracy and then corrected for insensible losses.

The reputation of the balance technique has been harmed by a few studies with inadequate equilibration times and short collection periods, but this should not be allowed to detract from the value of the meticulous work of those who have collected faecal and urinary samples for weeks or months from subjects on well-defined diets. This meticulous work has produced valuable balance data, which are clearly valid; the mean duration of the balances in the 210 studies from eight publications used in this report was 90 days with a range of 6-480 days. (The four 6-day balances in the series used a non-absorbable marker and are therefore acceptable.)

The usual way of determining mean calcium requirement from balance studies has been by linear regression of calcium output (or calcium balance) on intake and calculation of the mean intake at which intake and output are equal (or balance is sero). This was probably first done in 1939 by Mitchell and Curzon, who arrived at a mean requirement of 9. 8 mg/kg/ day or about 640 mg (16 mmol) at a mean body weight of 65 kg. The same type of calculation was subsequently used by many other workers who arrived at requirements ranging from 200 mg/day (5 mmol/day) in male Peruvian prisoners to 990 mg (24. 75 mmol) in premenopausal women, but most values were about 600 mg (15 mmol) without allowing for insensible losses.

However, this type of simple linear regression yields a higher mean calcium requirement (640 mg in the same 210 balances) than the intercept of absorbed and excreted calcium (520 mg) because it tends to underestimate the negative calcium balance at low intake and overestimate the positive balance at high intake. A better reflection of biological reality is obtained by deriving calcium output from the functions given in the previous section and then regressing that output on calcium intake. This yields where the negative balance is more severe at low intakes and less positive at high intakes than in the linear model and in which sero balance occurs at 520 mg.

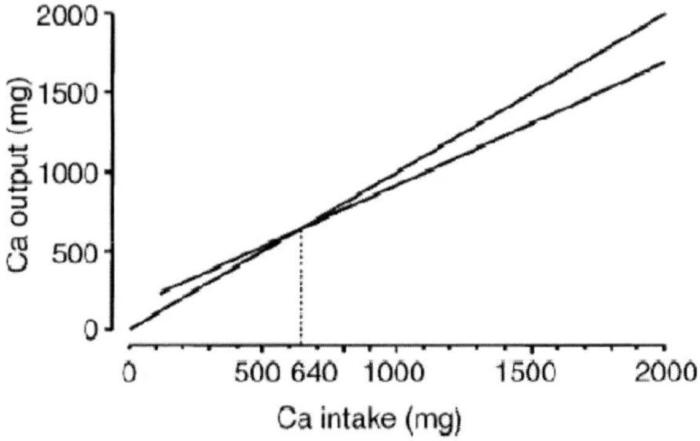

Fig. Calcium output as a linear (A) function of calcium intake calculated from the same balances as Figure

The equation is: $Ca_o = 0.779\ Ca_i + 142$ mg where Ca_o is calcium output.

An alternative way of calculating calcium requirement is to determine the intake at which the mean maximum positive balance occurs. This has been done with a two-component, split, linear regression model in which calcium balance is regressed on intake to determine the threshold intake above which no further increase in calcium retention occurs. This may well be an appropriate way of calculating the calcium requirement of children and adolescents (and perhaps pregnant and lactating women) who need to be in positive calcium balance and in whom the difference between calcium intake and output is therefore relatively large and measurable by the balance technique. However, in normal adults the difference between calcium intake and output at high calcium intakes represents a very small difference between two large numbers, and this calculation therefore carries too great an error to calculate their requirement.

We are inclined to think that the most satisfactory way of calculating calcium requirement from current data is as the intake at which excreted calcium equals net absorbed calcium, which has the advantage of permitting separate analysis of the effects of changes in calcium absorption and excretion. This intercept has been shown in Figure to occur at an intake of about 520 mg, but when insensible losses of calcium of 60 mg (1. 5 mmol) are taken into account, the intercept rises to 840 mg, which we believe is as close as it is possible to get at present to the calcium requirement of adults on Western-style diets.

The addition to this excretion line of an additional obligatory urinary calcium of 30 mg (0. 75 mmol) at menopause (50) raises the amount to about 1100 mg, which we suggest is the mean calcium requirement of postmenopausal women. However, this type of calculation cannot easily be applied to other high-risk populations (such as children) because there are not sufficient published data from these groups to permit a similar analysis of the relationship among calcium intake, absorption, and excretion. An alternative is to estimate how much calcium each population group needs to absorb to meet obligatory calcium losses and desirable calcium retention and then to calculate the intake required to provide this rate of calcium absorption. This is what has been done in the following section.

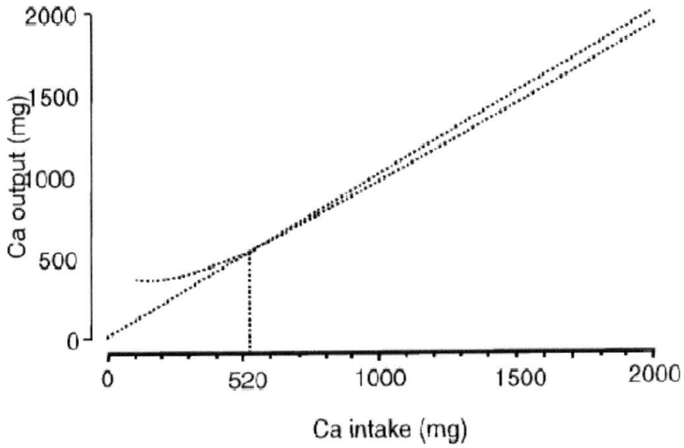

Fig. Calcium output as a non-linear function of calcium intake calculated from the same balances as Figure

Populations at Risk

It is clear from Figure that a positive calcium balance (i. e., net calcium retention) is required throughout growth, particularly during the first 2 years of life and during puberty and adolescence. These age groups therefore constitute populations at risk for calcium deficiency, as are pregnant women (especially in the last trimester), lactating women, postmenopausal women, and, possibly, elderly men. Our calculations for these groups, ultimately derived from Western European and North American data, are given below.

Recommendations by Group

Infancy

In the first 2 years of life, the daily calcium increment in the skeleton is about 100 mg (2. 5 mmol) (51). The urinary calcium of infants is about 10 mg/day (0. 25 mmol/day) and is virtually independent of intake (52-56) and insensible losses are perhaps the same. Therefore, infants need to absorb some 120 mg (3 mmol) of calcium daily to allow for normal growth. What this represents in dietary terms can be calculated from calcium absorption studies in newborn infants (52-56), which suggest that the absorption of calcium from cow milk by infants is about 0. 5 SD above the normal adult slope and from human milk is more than 1 SD above the normal adult slope.

If this information is correct, different recommendations need to be made for infants depending on milk source. With human milk, an absorption of 120 mg (3 mmol) of calcium requires a mean intake of 240 mg (6 mmol) and a recommended intake of say 300 mg (7. 5 mmol), which is close to the amount provided in the average daily milk production of 750 ml. With cow milk, calcium intake needs to be about 300 mg (7. 5 mmol) to meet the requirement and the allowance should be 400 mg (10 mmol).

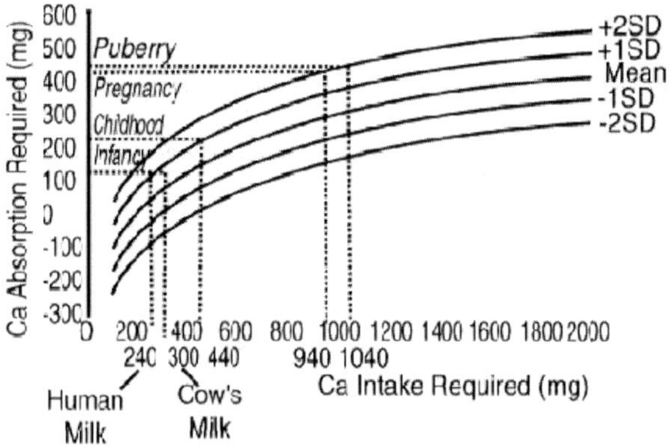

Fig. Calcium intakes required to provide the absorbed calcium necessary to meet calcium requirements at different stages in the lifecycle

Childhood

The accumulation of whole-body calcium with skeletal growth is illustrated in Figure. It rises from about 120 g (3 mol) at age 2 years to 400 g (10 mol) at age 9 years. These values can be converted into a daily rate of calcium accumulation from ages 2 to 9 of about 120 mg (3 mmol), which is very similar to the amount calculated by Leitch and Aitken (57) from growth analyses. Although urinary calcium must rise with the growth-related rise in glomerular filtration rate, a reasonable estimate of the mean value from ages 2 to 9 might be 60 mg (1. 5 mmol) (58).

When this is added to a daily skeletal increment of 120 mg (3 mmol) and a dermal loss of perhaps 40 mg (1. 0 mmol), the average daily net absorbed calcium needs to be 220 mg (5. 5 mmol) during this period. If the net absorption of calcium by children is 1 SD above that of adults, the average daily requirement during this period is about 440 mg (11 mmol) and the average recommended intake is 600 mg (15 mmol) - somewhat lower in the earlier years and somewhat higher in the later years.

Table: Recommended calcium allowances based on Western European, American and Canadian data

Group	Recommended intake mg/day
Infants and children	
0-6 months	
Human milk	300
Cow milk	400
7-12 months	400
1-3 years	500
4-6 years	600
7-9 years	700
Adolescents, 10-18 years	1300[a]
Adults	
Females	
19 years to menopause	1000
Postmenopause	1300
Males	
19-65 years	1000
65 +	1300
Pregnancy (last trimester)	1200
Lactation	1000

Puberty and adolescence

As can be seen in Figure, a striking increase in the rate of skeletal calcium accretion occurs at puberty - from about ages 10 to 17 years. The peak rate of calcium retention in this period is 300-400 mg (7. 5-10 mmol) daily (57); it occurs earlier in girls but continues longer in boys. For a target value of 300 mg (7. 5 mmol) for the skeleton, 100 mg (2. 5 mmol) for urinary calcium (58), and insensible losses of 40 mg (1. 0 mmol), the net absorbed calcium during at least part of this period needs to be 440 mg (11 mmol) daily. Even with assuming high calcium absorption (+2 SD), this requires an intake of 1040 mg (26. 0 mmol) daily and an allowance of 1300 mg (32. 5 mmol) during the peak growth phase. It is difficult to justify any difference between the allowances for boys and girls because, as mentioned above, although the growth spurt starts earlier in girls, it continues longer in boys. This recommended intake (which is close to that derived differently by Matkovic and Heaney [49,58]) is not achieved by many adolescents even in developed countries (59-61), but the effects of this shortfall on their growth and bone status are unknown.

Adults

As indicated earlier and for the reasons given, we accept that the mean apparent calcium requirement of adults in developed countries is about 520 mg (13 mmol) but that this is increased by insensible losses to some 840 mg (21 mmol). This reasoning forms the basis of our recommended intake for adults of 1000 mg.

Menopause

The most important single cause of osteoporosis - at least in developed countries - is probably menopause, which is accompanied by an unequivocal and sustained rise in obligatory urinary calcium of about 30 mg (0. 75 mmol) daily. Because calcium absorption certainly does not increase at this time - and probably decreases this extra urinary calcium represents a negative calcium balance which is compatible with the average bone loss of about 0. 5-1. 0 percent per year after menopause. There is a consensus that these events are associated with an increase in bone resorption but controversy continues about whether this is the primary event, the response to an increased calcium demand, or both. The results of calcium trials are clearly relevant.

Before 1997, there had been 20 prospective trials of calcium supplementation in 857 postmenopausal women and 625 control subjects; these trials showed highly significant suppression of bone loss by calcium. Another meta-analysis covering similar numbers showed that calcium supplementation significantly enhanced the effect of oestrogen on bone. It is therefore logical to recommend sufficient additional calcium after the menopause to cover at least the extra obligatory loss of calcium in the urine. The additional dietary calcium needed to meet an increased urinary loss of 30 mg (0. 75 mmol) is 260 mg/day (6. 5 mmol/day), which raises the daily requirement from 840 mg (21 mmol) to 1100 mg (27. 5 mmol) and the recommended intake from 1000 to 1300 mg/day (25 to 32. 5 mmol/day), which is a little higher than that recommended by the United States and Canada.

Ageing

Not enough is known about bone and calcium metabolism during ageing to enable calculation of the calcium requirements of older men and women with any confidence. Calcium absorption tends to decrease with age in both sexes but whereas the evidence that calcium requirement goes up at the menopause is strong, corresponding evidence about ageing men is less convincing. Nonetheless, as a precaution we propose an extra allowance of 300 mg/day (7. 5 mmol/day) for men over 65 to bring them into line with postmenopausal women.

Pregnancy

The calcium content of the newborn infant is about 24 g (600 mmol). Most of this calcium is laid down in the last trimester of pregnancy, during which the foetus retains about 240 mg (6 mmol) of calcium daily (51). There is some evidence that pregnancy is associated with an increase in calcium absorption [associated with a rise in the plasma 1,25$(OH)_2$ D level] (70-72). For a maternal urinary calcium of 120 mg (3 mmol) and a maternal skin loss of 60 mg (1. 5 mmol), the absorbed calcium should be 420 mg (10. 5 mmol) daily. Even at optimal calcium absorption, the corresponding calcium intake would need to be 940 mg (23. 5 mmol) and the recommended allowance would need to be 1200 mg (30 mmol).

Lactation

The calcium content of human milk is about 36 mg per 100 ml (9 mmol/l). A lactating woman produces about 750 ml of milk daily, which represents about 280 mg (7. 0 mmol) of calcium. For a maternal urinary calcium of 100 mg/day (2. 5 mmol/day) and maternal skin loss of 60 mg/day the required absorption is 440 mg/day (11 mmol/day) - the same as at puberty. If calcium absorption efficiency is maximal (i. e., 2 SD above the normal mean) - possibly because of the effect of prolactin on the production of 1,25$(OH)_2$D the requirement would be about 1040 mg (26. 0 mmol) and the recommended intake would be about 1300 mg (32. 5 mmol).

However, although it is known that bone is lost during lactation and restored after weaning, early reports that this bone loss could be prevented by calcium supplementation have not been confirmed. The prevailing view now is that calcium absorption does not increase and may decrease during lactation. It is increasingly thought that lactational bone loss is not a nutritional problem but may be due to parathyroid hormone-related peptide secreted by the breast and therefore beyond the control of dietary calcium. In view of this uncertainty, we do not at present recommend any extra calcium allowance during lactation; any risk to adolescent mothers is covered by our general recommendation of 1300 mg for adolescents.

Upper Limits

Because of the inverse relationship between fractional calcium absorption and calcium intake, a calcium supplement of 1000 mg (2. 5 mmol) added to a Western-style diet only increases urinary calcium by about 60 mg (1. 5 mmol). Urinary calcium also rises very slowly with intake (slope of 5-10 percent) and the risk of kidney stones from dietary hypercalciuria must therefore be negligible. In fact, it has been suggested that dietary calcium may protect against renal calculi because it binds dietary oxalate and reduces oxalate excretion.

Toxic effects of a high calcium intake have only been described when the calcium is given as the carbonate in very high doses; this toxicity is caused as much by the alkali as by the calcium and is due to precipitation of calcium salts in renal tissue (milk-alkali syndrome). However, in practice we recommend an upper limit on calcium intake of 3 g (75 mmol).

Comparisons with Other Recommendations

Our recommendations in Table can be compared with the current recommendations of Australia, the United Kingdom, the European Union, and the United States and Canada in Table. Our recommendations for adults are very close to those of the United States and Canada but higher than those of the United Kingdom and Australia, which do not take into account insensible losses, and higher than those of the European Union, which assumed 30 percent absorption of dietary calcium.

The British and European values make no allowance for ageing or menopause. Recommendations for other high-risk groups are very similar in all five sets of recommendations except for the rather low allowance for infants by the United States and Canada. Nonetheless, and despite this broad measure of agreement, we have some misgivings about the application of these recommendations, all of which rely ultimately on data from developed nations, to developing countries where other dietary constituents - such as animal protein and sodium - and environmental factors may be very different. We shall therefore in the next sections briefly review current knowledge about the prevalence of osteoporosis across racial national boundaries and its relevance to calcium requirement.

Intakes of calcium have been known for many years to vary greatly from one country to another, as is clearly shown in FAO food balance sheets. Until fairly recently, it was widely assumed that low calcium intakes had no injurious consequences. This view of the global situation underlay the very conservative adequate calcium intakes recommended by WHO/FAO. At that time, osteoporosis was still regarded as a bone matrix disorder and the possibility that it could be caused by calcium deficiency was barely considered.

The paradigm has changed since then. Calcium deficiency is taken more seriously than it was and the apparent discrepancy between calcium intake and bone status across the world has attracted more attention. In general, recent investigations have sought for evidence of low bone density and high fracture incidence in countries where calcium intake is low; rickets has not been looked for, but the low calcium rickets recently reported from Nigeria will no doubt attract attention.

This issue can be considered at several levels. The first level is genetic: Is there a genetic (ethnic) difference in the prevalence of osteoporosis between racial groups within a given society? The second level might be termed environmental-cultural (e. g., dietary): Is there a difference in the prevalence of osteoporosis between national groups of similar ethnic composition? The third level is environmental-geographical (e. g., latitude, affluence, and lifestyle): Is there a difference in the prevalence of osteoporosis between countries regardless of ethnic composition? At each of these levels, the prevalence of osteoporosis can in theory be determined in at least two ways - from the distribution of bone density within the population and from the prevalence of fractures, notably hip fractures. In practice, hip fracture data (or mortality from falls for elderly people which has been used as a surrogate [84]) are more readily available than bone densitometry.

Ethnicity, Environment, and Lifestyle

The conclusion from the above is that there are probably ethnic differences in hip fracture rates within countries but also environmental differences within the same ethnic group which may complicate the story. For international comparisons on a larger scale, it is impossible to separate genetic from environmental factors, but certain patterns emerge which are likely to have biological meaning. The most striking of these is the positive correlation between hip fracture rates and standard of living noted by Hegsted when he observed that osteoporosis was largely a disease of affluent Western cultures. He based this conclusion on a previously published review of hip fracture rates in 10 countries, which strongly suggested a correlation between hip fracture rate and affluence. Another review of 19 regions and racial groups confirmed this by showing a gradient of age- and sex-adjusted hip fracture rates from 31 per 100 000 in South African Bantu to 968 per 100 000 in Norway. In the analysis of hip fracture rates in Beijing and Hong Kong referred to above, it was noted that the rates in both cities were much lower than in the United States. Many other publications point to the same conclusion - that hip fracture prevalence (and by implication osteoporosis) is related to affluence and, consequently, to animal protein intake, as Hegsted pointed out, but also and paradoxically to calcium intake.

The Calcium Paradox

The paradox that hip fracture rates are higher in developed nations where calcium intake is high than in developing nations where calcium intake is low clearly calls for an explanation. Hegsted was probably the first to note the close relation between calcium and protein intakes across the world and to hint at but dismiss the possibility that the adverse effect of protein might outweigh the positive effect of calcium on calcium balance. Only recently has fracture risk been shown to be a function of protein intake in American women. There is also suggestive evidence that hip fracture rates (as judged by mortality from falls in elderly people across the world) are a function of protein intake, national income, and latitude. The latter is particularly interesting in view of the strong evidence of vitamin D deficiency in hip fracture patients in the developed world and the successful prevention of such fractures with small doses of vitamin D and calcium. It is therefore possible that hip fracture rates may be related to protein intake, vitamin D status, or both and that either of these factors could explain the calcium paradox. We shall therefore consider how these and other nutrients (notably sodium) affect calcium requirement.

Nutritional Factors Affecting Calcium Requirements

The calculations of calcium requirements proposed above were based on data from developed countries (notably the United States and Norway) and can only be applied with any confidence to nations and populations with similar dietary cultures. Other dietary cultures may entail different calcium requirements and call for different recommendations. In particular, the removal or addition of any nutrient that affects calcium absorption or excretion must have an effect on calcium requirement. Two such nutrients are sodium and animal protein, both of which increase urinary calcium and must be presumed therefore to increase calcium requirement. A third candidate is vitamin D because of its role in calcium homeostasis and calcium absorption.

Sodium

It has been known at least since 1961 that urinary calcium is related to urinary sodium and that sodium administration raises calcium excretion, presumably because sodium competes with calcium for reabsorption in the renal tubules. Regarding the quantitative relationships between the renal handling of sodium and calcium, the filtered load of sodium is about 100 times that of calcium (in molar terms) but the clearance of these two elements is similar at about 1 ml/min, which yields about 99 percent reabsorption and 1 percent excretion for both. However, these are approximations, which conceal the close dependence of urinary sodium on sodium intake and the weaker dependence of urinary calcium on calcium intake. It is an empirical fact that urinary sodium and calcium are significantly related in normal and hypercalciuric subjects on freely chosen diets. The slope of urinary calcium on sodium varies in published work from about 0. 6 percent to 1. 2 percent (in molar terms); a representative figure is about 1 percent - that is, 100 mmol of sodium (2. 3 g) takes out about 1 mmol (40 mg) of calcium.

The biological significance of this relationship is supported by the accelerated osteoporosis induced by feeding salt to rats on low-calcium diets and the effects of salt administration and salt restriction on markers of bone resorption in postmenopausal women. Because salt restriction lowers urinary calcium, it is likely also to lower calcium requirement and, conversely, salt feeding is likely to increase calcium requirement. Lowering sodium intake by 100 mmol (2. 3 g) from, for example, 150 to 50 mmol (3. 45 to 1. 15 g), reduces the theoretical calcium requirement from 840 mg (21 mmol) to 600 mg (15 mmol). However, the implications of this on calcium requirement across the world cannot be computed because information about sodium intakes is available from very few countries.

Protein

The positive effect of dietary protein - particularly animal protein - on urinary calcium has also been known at least since the 1960s. One study found that 0. 85 mg of calcium was lost for each gram of protein in the diet (130). A meta-analysis of 16 studies in 154 adult humans on protein intakes up to 200 g found that 1. 2 mg of calcium was lost in the urine for every 1g rise in dietary protein (131). A small but more focussed study showed a rise of 40 mg in urinary calcium when dietary animal protein was raised from 40 to 80 g (i. e., within the physiological range) (132). This ratio of urinary calcium to dietary protein ratio (1mg to 1g) is a representative value, which we have adopted. This means that a 40g reduction in animal protein intake from 60 to 20 g (or from the developed to the developing world would reduce calcium requirement by the same amount as a 2. 3g reduction in dietary sodium, i. e. from 840 to 600 mg.

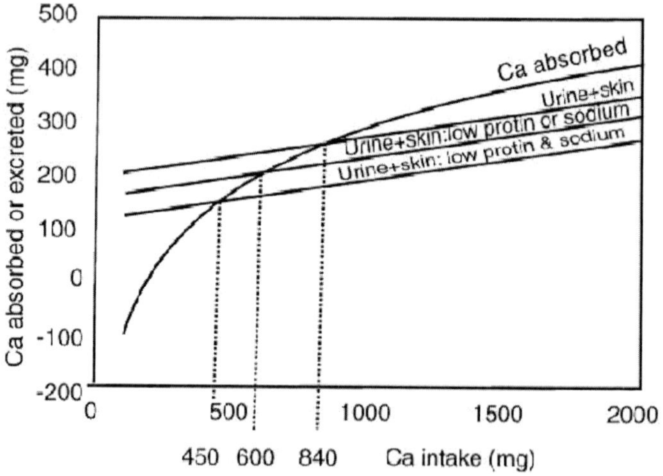

Fig. The effect of varying protein or sodium intake on theoretical calcium requirement.

How animal protein exerts its effect on calcium excretion is not fully understood. A rise in glomerular filtration rate in response to protein has been suggested as one factor but this is unlikely to be important in the steady state. The major mechanisms are thought to be the effect of the acid load contained in animal proteins and the complexing of calcium in the renal tubules by sulphate and phosphate ions released by protein metabolism. Urinary calcium is significantly related to urinary phosphate (as well as to urinary sodium), particularly in subjects on restricted calcium intakes or in the fasting state, and most of the phosphorus in the urine of people on Western-style diets comes from animal protein in the diet. Similar considerations apply to urinary sulphate but it is probably less important than the phosphate ion because the association constant for calcium sulphate is lower than that for calcium phosphate. The empirical observation that each 1 g of protein results in 1 mg of calcium in the urine agrees very well with the phosphorus content of animal protein (about 1 percent by weight) and the observed relationship between calcium and phosphate in the urine.

Vitamin D

One of the first observations made on vitamin D after it had been identified was that it promoted calcium absorption. It is now well established that vitamin D (synthesised in the skin under the influence of sunlight) is converted to 25OHD in the liver and then to 1,25$(OH)_2$D in the kidneys and that the latter metabolite controls calcium absorption. However, plasma 25OHD closely reflects vitamin D nutritional status and because it is the substrate for the renal enzyme which produces 1,25$(OH)_2$D, it could have an indirect effect on calcium absorption.

The plasma level of 1,25-$(OH)_2$D is principally regulated is through increased gene expression of the 1-a-hydroxylase (CYP1a) and not by increased 25OHD levels. This has been seen consistently in animal studies, and the high calcium absorption and high plasma 1,25-(OH) $_2$D observed in Gambian mothers is consistent with this type of adaptation. However, increasing latitude may compromise vitamin D synthesis to the degree that 25OHD levels are no longer sufficient to sustain adequate 1,25-$(OH)_2$D levels and efficient intestinal calcium absorption, although this theory remains unproved. Regardless of the mechanism of compromised vitamin D homeostasis, the differences in calcium absorption efficiency have a major effect on theoretical calcium requirement, which shows that an increase in calcium absorption of as little as 10 percent reduces the intercept of excreted and absorbed calcium from 840 to 680 mg.

Implications

The major reduction in theoretical calcium requirement which follows animal protein restriction has led us to attempt to show in Table how the calcium allowances recommended in Table could be modified to apply to nations where the animal protein intake per capita is around 20-40 g rather than around the 60-80 g in developed countries. These hypothetical allowances take into account the need to protect children, in whom skeletal needs are much more important determinants of calcium requirement than are urinary losses and in whom calcium supplementation had a beneficial effect in the Gambia. However, adjustment for animal protein intake has a major effect on the recommended calcium allowances for adults as the table shows. It also brings the allowances nearer to what the actual calcium intakes are in many parts of the world.

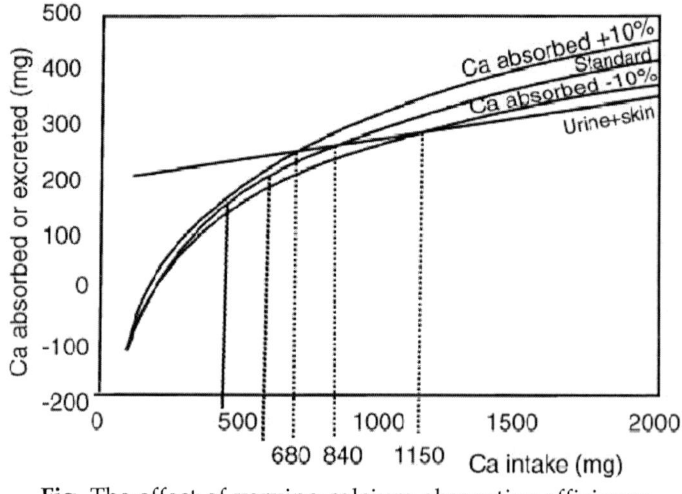

Fig. The effect of varying calcium absorptive efficiency on theoretical calcium requirement

If sodium intakes were also lower in developing than developed nations or urinary sodium were reduced for other reasons such as increased sweat losses, the calcium requirement might be even lower, for example, 450 mg. This would be reduced still further by any increase in calcium absorption, whether resulting from better vitamin D status because of increased sunlight exposure or for other reasons. Because the increase in calcium absorption in the Gambia is much more than 10 percent, this is likely to have a major - although not at present calculable - effect on calcium requirement there. However, the adjusted bone mineral density in Gambian women is reported to be some 20 percent lower in the spine (but not in the forearm) than in British women, which emphasises the need for more data from developing countries.

Table: Theoretical calcium allowances based on an animal protein intake of 20-40 g

Group	Recommended intake mg/day
Infants and children	
0-6 months	
Human milk	300
Cow milk	400
7-12 months	450
1-3 years	500
4-6 years	550
7-9 years	700
Adolescents, 10-18 years	1000[a]
Adults	
Females 19 years to menopause	750
Postmenopause	800
Males	
19-65 years	750
65 +	800
Pregnancy (last trimester)	800
Lactation	750

[a] Particularly during the growth spurt.

Conclusions

Calcium is an essential nutrient that plays a vital role in neuromuscular function, many enzyme-mediated processes, blood clotting, and providing rigidity to the skeleton by virtue of its phosphate salts. Its non-structural roles require the strict maintenance of ionised calcium concentration in tissue fluids at the expense of the skeleton if necessary and it is therefore the skeleton which is at risk if the supply of calcium falls short of requirement.

Calcium requirements are essentially determined by the relationship between absorptive efficiency and excretory rate - excretion being through the bowel, kidneys, skin, hair, and nails. In adults, the rate of calcium absorption from the gastrointestinal tract needs to match the rate of all losses from the body if the skeleton is to be preserved; in children and adolescents, an extra input is needed to cover the requirements of skeletal growth.

Compared with that of other minerals, calcium economy is relatively inefficient. On most intakes, only about 25-30 percent of dietary calcium is effectively absorbed and obligatory calcium losses are relatively large. Absorbed calcium has to match these obligatory losses and the dietary intake has to be large enough to ensure this rate of absorption if skeletal damage is to be avoided. The system is subject to considerable inter-individual variation in both calcium absorption and excretion for reasons that are not fully understood but which include vitamin D status, sodium and protein intake, age, and menopausal status in women.

Although it needs to be emphasised that calcium deficiency and negative calcium balance must sooner or later lead to osteoporosis, this does not mean that all osteoporosis can be attributed to calcium deficiency. On the contrary, there may be more osteoporosis in the world from other causes. Nonetheless, it would probably be generally agreed that any form of osteoporosis must inevitably be aggravated by negative external calcium balance. Such negative balance - even for short periods - is prejudicial because it takes so much longer to rebuild bone than to destroy it. Bone that is lost, even during short periods of calcium deficiency, is only slowly replaced when adequate amounts of calcium become available.

In seeking to define advisable calcium intakes on the basis of physiologic studies and clinical observations, nutrition authorities have to rely largely on data from developed nations living at relatively high latitudes. Although it is now possible to formulate recommendations that are appropriate to different stages in the life cycle of the populations of these nations, extrapolation from these figures to other cultures and nutritional environments can only be tentative and must rely on what is known of nutritional and environmental effects on calcium absorption and excretion. Nonetheless, we have made an attempt in this direction, knowing that our speculative calculations may be incorrect because of other variables not yet identified.

No reference has been made in this account to the possible beneficial effects of calcium in the prevention or treatment of pre-eclampsia, colon cancer, or hypertension and no attempt has been made to use these conditions as endpoints on which to base calcium intakes. In each of the above conditions, epidemiologic data suggested an association with calcium intake, and experimentation with increased calcium intakes has now been tried. In each case the results have been disappointing, inconclusive, or negative and have stirred controversy.

Because there is no clear consensus about optimal calcium intake for prevention or treatment of these conditions and also no clear mechanistic ideas on how dietary calcium intakes affect them, it is not possible to allow for the effect of health outcomes in these areas on our calcium recom-mendations. However, although the anecdotal information and positive effects of calcium observed in these three conditions cannot influence our recommendations, they do suggest that generous calcium allowances may confer other benefits besides protecting the skeleton. Similarly, no reference has been made to the effects of physical activity, alcohol, smoking, or other known risk factors on bone status because the effects of these variables on calcium requirement are beyond the realm of simple calculation.

IODINE

Summary of the Human Metabolic Processes Requiring Iodine

At present, the only physiologic role known for iodine in the human body is in the synthesis of thyroid hormones by the thyroid gland. Therefore, the dietary requirement of iodine is determined by normal thyroxine (T_4) production by the thyroid gland without stressing the thyroid iodide trapping mechanism or raising thyroid stimulating hormone (TSH) levels.

Iodine from the diet is absorbed throughout the gastrointestinal tract. Dietary iodine is converted into the iodide ion before it is absorbed. The iodide ion is bio-available and absorbed totally from food and water. This is not true for iodine within thyroid hormones ingested for therapeutic purposes. Iodine enters the circulation as plasma inorganic iodide, which is cleared from circulation by the thyroid and kidney. The iodide is used by the thyroid gland for synthesis of thyroid hormones, and the kidney excretes iodine with urine. The excretion of iodine in the urine is a good measure of iodine intake. In a normal population with no evidence of clinical iodine deficiency either in the form of endemic goitre or endemic cretinism, urinary iodine excretion reflects the average daily iodine requirement. Therefore, for determining the iodine requirements, the important indexes are serum T_4 and TSH levels (indicating normal thyroid status) and urinary iodine excretion. The simplified diagram of metabolic circuit of iodine is given in Figure.

Overview of Significant Scientific Information

All biologic actions of iodide are attributed to the thyroid hormones. The major thyroid hormone secreted by the thyroid gland is T_4 (tetra-iodo-thyronine). T_4 in circulation is taken up by the cells and is de-iodinated by the enzyme 5'prime-mono-de-iodinase in the cytoplasm to convert it into tri-iodo-thyronine (T_3), the active form of thyroid hormone. T_3 traverses to the nucleus and binds to the nuclear receptor. All the biologic actions of T_3 are mediated through the binding to the nuclear receptor, which controls the transcription of a particular gene to bring about the synthesis of a specific protein.

The physiologic actions of thyroid hormones can be categorised as 1) growth and development and 2) control of metabolic processes in the body. Thyroid hormones play a major role in the growth and development of brain and central nervous systems in humans from the 15th week of gestation to age 3 years. If iodine deficiency exists during this period and results in thyroid hormone deficiency, the consequence is derangement in the development of brain and central nervous system. These derangements are irreversible, the most serious form being that of cretinism. The effect of iodine deficiency at different stages of life is given in Table.

The other physiologic role of thyroid hormone is to control several metabolic processes in the body. These include carbohydrate, fat, protein, vitamin, and mineral metabolism. For example, thyroid hormone increases energy production, increases lipolysis, and regulates neoglucogenesis, and glycolysis.

Table: The spectrum of iodine deficiency disorders

Life stage Stage in life	Effects
Foetus	Abortions Stillbirths Congenital anomalies Increased perinatal mortality Increased infant mortality Neurological cretinism: mental deficiency, deaf mutism, spastic diplegia, and squint Myxedematous cretinism: mental deficiency and dwarfism Psychomotor defects
Neonate	Neonatal goitre Neonatal hypothyroidism
Child and Adolescent	Goitre Juvenile hypothyroidism Impaired mental function Retarded physical development
Adult	Goitre with its complications Hypothyroidism Impaired mental function

Population at risk

Iodine deficiency affects all stages of human life, from the intra-uterine stage to old age, as shown in Table. However, pregnant women, lactating women, women of reproductive age, and children younger than 3 years are considered to be at high risk. During foetal and neonatal growth and development, iodine deficiency leads to irreversible damage to the brain and central nervous system.

Dietary Sources

The iodine content of food depends on the iodine content of the soil in which it is grown. The iodine present in the upper crust of earth is leached by glaciation and repeated flooding and is carried to the sea. Sea water is, therefore, a rich source of iodine. The seaweed located near coral reefs has an inherent biologic capacity to concentrate iodine from the sea. The reef fish which thrive on seaweed are rich in iodine. Thus, a population consuming seaweed and reef fish has a high intake of iodine, as the case in Japan.

The amount of iodine intake by the Japanese is in the range of 2-3 mg/day. In several areas of Asia, Africa, Latin America, and parts of Europe, iodine intake varies from 20 to 80 mg/day. In the United States and Canada and some parts of Europe, the intake is around 500 mg/day.

Table: Average iodine content of foods (in ìg/g)

Food	Fresh basis		Dry basis	
	Mean	Range	Mean	Range
Fish (fresh water)	30	17-40	116	68-194
Fish (marine)	832	163-3180	3715	471-4591
Shellfish	798	308-1300	3866	1292-4987
Meat	50	27-97	-	-
Milk	47	35-56	-	-
Eggs	93	-	-	-
Cereal grains	47	22-72	65	34-92
Fruits	18	10-29	154	62-277
Legumes	30	23-36	234	223-245
Vegetables	29	12-201	385	204-1636

The iodine content of food varies with geographic location because there is a large variation in the iodine content of the inorganic world. Thus, the average iodine content of foods.

Recommended intake

The daily intake of iodine recommended by the National Research Council of the US National Academy of Sciences in 1989 was 40 μg/day for young infants (0-6 months), 50 μg/day for older infants (6-12 months), 60-100 μg/day for children (1-10 years), and 150 μg/day for adolescents and adults. These values approximate 7. 5 μg/kg/day for age 0-12 months, 5. 4 μg/kg/day for age 1-10 years, and 2 μg/kg/day for adolescents and adults. These amounts are proposed to allow normal T_4 production without stressing the thyroid iodide trapping mechanism or raising TSH levels.

Iodine Requirements in Infancy

An average human-milk intake of 750 ml/day would give an intake of iodine of about 60 μg/day in Europe and 120 μg/day in the United States. The upper US value (490 μg/l) would provide 368 μg/day or 68 μg/kg/day for a 5-kg infant. Positive iodine balance in the young infant, which is required for the increasing iodine stores of the thyroid, is achieved only when the iodine intake is at least 15 μg/kg/day in full-term infants and 30 μg/kg/day in pre-term infants. The iodine requirement of pre-term infants is twice that of term infants because of a 50 percent lower retention of iodine by pre-term infants. This corresponds approximately to an iodine intake of 90 μg/day. (This is probably based on the assumption of average body weight of 6 kg for a child of 6 months, the mid-age of an infant.) This value is twofold higher than the US recommendations.

On the basis of these considerations, a revision is proposed for the earlier World Health Organization (WHO), United Nations Children's Fund (UNICEF), and International Council for the Control of Iodine Deficiency Disorders (ICCIDD) recommendations: an iodine intake of 90 μg/day from birth onwards is suggested. To reach this objective, and based on an intake of milk of about 150 ml/kg/day, the iodine content of formula milk should be increased from 50 to 100 μg/l for full-term infants and to 200 μg/l for pre-term infants.

Table: Iodine content of the inorganic world

Location	Iodine content
Terrestrial air	1. 0 μg/l
Marine air	100. 0 μg/l
Terrestrial water	5. 0 μg/l
Sea water	50. 0 μg/l
Igneous rocks	500. 0 μg/kg
Soils from igneous rocks	9000. 0 μg/kg
Sedimentary rocks	1500. 0 μg/kg
Soils from sedimentary rocks	4000. 0 μg/kg
Metamorphic rocks	1600. 0 μg/kg
Soils from the metamorphic rocks	5000. 0 μg/kg

Iodine Requirements in Children

The daily iodine need on a body weight basis decreases progressively with age. A study by Tovar and colleagues correlating 24-hour thyroid radioiodine uptake and urinary iodine excretion in 9-13-year-old schoolchildren in rural Mexico suggested that an iodine intake in excess of 60 μg/day is associated with a 24-hour thyroidal radioiodine uptake below 30 percent. Lower excretion values are associated with higher uptake values. This would approximate 3 μg/kg/day in an average sise 10-year-old child (approximate body weight of 20 kg), so that an intake of 60-100 μg/day for child of 1-10 years seems appropriate. These requirements are based on the body weight of Mexican children who participated in this study. The average body weight of a 10-year-old child, as per the Food and Agriculture Organization references, is 25 kg. Thus, the iodine requirement for a 1-10-year-old child would be 90-120 ìg/day.

Iodine requirements in adults

Iodine at 150 μg/day for adolescents and adults is justified by the fact that it corresponds to the daily urinary excretion of iodine and to the iodine content of food in non-endemic areas (areas where iodine intake is adequate). It also provides the iodine intake necessary to maintain the plasma iodide level above the critical limit of 0. 10 μg/dl, which is the average level likely to be associated with the onset of goitre. Moreover, this level of iodine intake is required to maintain the iodine stores of the thyroid above the critical threshold of 10 mg, below which an insufficient level of iodisation of thyroglobulin leads to disorders in thyroid hormone synthesis.

Data reflecting either iodine balance or its effect on thyroid physiology can help to define optimal iodine intake. In adults and adolescents in equilibrium with their nutritional environment, most dietary iodine eventually appears in the urine, so the urinary iodine concentration is a useful measure for assessing iodine intake. For this, casual samples are sufficient if enough are collected and if they accurately represent a community. A urinary iodine concentration of 100 μg/L corresponds to an intake of about 150 μg/day in the adult. Median urinary iodine concentrations below 100 μg/l in a population are associated with increases in median thyroid sise and in serum TSH and thyroglobulin values.

Correction of the iodine deficiency will bring all these measures back into the normal range. Recent data from the Thyro-Mobil project in Europe have confirmed these relations by showing that the largest thyroid sises are associated with the lowest urinary iodine concentrations. Once a median urinary iodine excretion of about 100 μg/L is reached, the ratio of thyroid sise to body sise remains fairly constant. A urinary iodine excretion between 151 and 200 μg/g creatinine (1. 18-1. 57 mmol/g creatinine), corresponding to a concentration of about 200 μg/l (1. 57 mmol/l), gave the lowest values for serum TSH in a non-goitrous population. Similar recent data from Australia show that the lowest serum TSH and thyroglobulin values were associated with urine containing 200-300 μg iodine/g creatinine.

Other investigations followed serum TSH levels in subjects without thyroid glands who were given graded doses of T_4 and found that euthyroidism established in adults with an average daily dose of 100 μg T_4 would require at least 65 μg of iodine with maximal efficiency of iodine use by the thyroid. In practice such maximal efficiency is never obtained and therefore considerably more iodine is necessary. Data from controlled observations associated a low urinary iodine concentration with a high goitre prevalence, high radioiodine uptake, and low thyroidal organic iodine content. Each of these measures reached a steady state once the urinary iodine excretion was 100 μg/l (0. 78 mmol/l)or greater.

Iodine requirements in pregnancy

The iodine requirement during pregnancy is increased to provide for the needs of the foetus and to compensate for the increased loss of iodine in the urine resulting from an increased renal clearance of iodine during pregnancy. These requirements have been derived from studies of thyroid function during pregnancy and in the neonate under conditions of moderate iodine deficiency.

For example, in Belgium, where the iodine intake is estimated to be 50-70 μg/day, thyroid function during pregnancy is characterised by a progressive decrease of the serum concentrations of thyroid hormones and an increase in serum TSH and thyroglobulin. Thyroid volume progressively increases and is above the upper limit of normal in 10 percent of the women by the end of pregnancy. Serum TSH and thyroglobulin are still higher in the neonates than in the mothers.

These abnormalities are prevented only when the mother receives a daily iodide supplementation of 161 μg/day during pregnancy (derived from 131 μg potassium iodide and 100 μg T_4 given daily). T_4 with iodine was probably administered to the pregnant women to rapidly correct sub-clinical hypothyroidism, which would not have occurred if iodine had been administered alone. These data indicate that the iodine intake required to prevent the onset of sub-clinical hypothyroidism of mother and foetus during pregnancy, and thus to prevent the possible risk of brain damage of the foetus, is approximately 200 μg/day.

On the basis of the considerations reviewed above for the respective population groups to meet the daily iodine requirements, revisions of the current recommendations for daily iodine intak.

Table: Proposed revision for daily iodine intake recommendations of 1996 by the World Health Organization, United Nations Children's Fund, and International Council for the Control of Iodine Deficiency Disorders

Population sub-groups	Total iodine intake μg/day	Iodine ìg/kg/day
Infants (first 12 months)	90[a]	15. 0
Children (1-6 years)	90	6. 0
Schoolchildren (7-12 years)	120	4. 0
Adults (12+ years)	150	2. 0
Pregnant and lactating women	200	3. 5

[a] Revised to 90 μg from the earlier recommendation of 50 μg.

Upper Limit of Iodine Intake for Different Age Groups

An iodine excess also can be harmful to the thyroid of infants by inhibiting the process of synthesis and release of thyroid hormones. The threshold upper limit of iodine intake (the intake beyond which thyroid function is inhibited) is not easy to define because it is affected by the level of iodine intake before exposure to iodine excess.

Indeed, long-standing moderate iodine deficiency is accompanied by an accelerated trapping of iodide and by a decrease in the iodine stores within the thyroid. Under these conditions, the critical ratio between iodide and total iodine within the thyroid, which is the starting point of the Wolff-Chaikoff effect, is more easily reached during iodine depletion than under normal conditions.

In addition, the neonatal thyroid is particularly sensitive to the Wolff-Chaikoff effect because the immature thyroid gland is unable to reduce the uptake of iodine from the plasma to compensate for increased iodine ingestion. For these reasons transient neonatal hypothyroidism or transient hyper TSHemia after iodine overload of the mother, especially after the use of povidone iodine, has been reported more frequently in European countries such as in Belgium, France, and Germany, which have prevailing moderate iodine deficiency.

Iodine Intake in Areas of Moderate Iodine Deficiency

In a study in Belgium, iodine overload of mothers (cutaneous povidone iodine) increased the milk iodine concentration and increased iodine excretion in the term newborns (mean weight about 3 kg). Mean milk iodine concentrations of 18 and 128 μg/dl were associated with average infant urinary iodine excretion levels of 280 and 1840 μg/l (2. 20-14. 48 mmol/l), respectively. Estimated average iodine intakes would be 112 and 736 μg/day, or 37 and 245 μg/kg/day, respectively.

The lower dose significantly increased the peak TSH response to exogenous thyroid releasing hormone but did not increase the (secretory) area under the TSH response curve. The larger dose increased both the peak response and secretory area as well as the baseline TSH concentration. Serum T_4 concentrations were not altered, however. Thus, these infants had a mild and transient, compensated hypothyroid state. Non-contaminated mothers secreted milk containing 9. 5 μg iodine/dl, and the mean urinary iodine concentration of their infants was 144 μg/l (1. 13 mmol/l). These data indicate that modest iodine overloading of term infants in the neonatal period in an area of relative dietary iodine deficiency also can impair thyroid hormone formation.

Similarly, studies in France indicated that premature infants exposed to cutaneous povidone iodine or fluorescinated alcohol-iodine solutions and excreting iodine in urine in excess of 100 μg/day manifested decreased T_4 and increased TSH concentrations in serum. The extent of these changes was more marked in premature infants with less than 34 weeks gestation than in those with 35-37 weeks gestation. The full-term infants were not affected. These studies suggest that in Europe the upper limit of iodine intake, which predisposes to blockage of thyroid secretion in premature infants (about 200 μg/day) is 2 to 3 times the average intake from human milk and about equivalent to the upper range of intake.

Iodine Intake in Areas of Iodine Sufficiency

Table summarises the recommended dietary intake of iodine for age and approximate level of intake which appear not to impair thyroid function in the European studies of Delange in infants, in the loading studies of adults in the United States, or during ingestion of the highest estimates of dietary intake (just reviewed) in the United States. Except for the values for premature infants, these probably safe limits are 15-20 times more than the recommended intakes. These data refer to all sources of iodine intake.

The average iodine content of infant formulas is approximately 5 μg/dl. The upper limit probably should be one that provides a daily iodine intake of no more than 100 μg/kg. For this limit and with the assumption that the total intake is from infant formula, with a daily intake of 150 ml/kg (100 kcal/kg), the upper limit of the iodine content of infant formula would be about 65 μg/dl. The current suggested upper limit of iodine in infant formulas of 75 μg/100 kcal (89μg/500 kJ or 50 μg/dl), therefore, seems reasonable.

Table: Recommended dietary intakes of iodine and probable safe upper limits

Group	Recommended μg/kg/day	Upper limit[a] μg/kg/day
Premature infants	30	100
Infants 0-6 months	15	150
Infants 7-12 months	15	140
Children 1-6 years	6	50
School children 7-12 years	4	50
Adolescents and adults (12+ years)	2	30
Pregnancy and lactation	3. 5	40

Excess Iodine Intake

Excess iodine intake is more difficult to define. Many people regularly ingest huge amounts of iodine - in the range 10-200 mg/day - without apparent adverse effects. Common sources are medicines (e. g., amiodarone contains 75 mg iodine per 200-mg capsule), foods (particularly dairy products), kelp (eaten in large amounts in Japan), and iodine-containing dyes (for radiologic procedures). Excess consumption of salt has never been documented to be responsible for excess iodine intake. Occasionally each of these may have significant thyroid effects, but generally they are tolerated without difficulty.

People without evidence of underlying thyroid disease almost always remain euthyroid in the face of large amounts of excess iodine and escape the acute inhibitory effects of excess intra-thyroidal iodide on the organification (i. e., attachment of ′oxidised iodine′ species to throsyl residues in the thyroid gland for the synthesis of thyroid hormones) of iodide and on subsequent hormone synthesis (escape from or adaptation to the acute Wolff-Chaikoff effect). This adaptation most likely involves a decrease in thyroid iodide trapping, perhaps corresponding to a decrease in the thyroid sodium-iodide transporter recently cloned. Some people, especially those with long-standing nodular goitre who live in iodine-deficient regions and are generally ages 40 years or older, may develop iodine-induced hyperthyroidism after ingestion of excess iodine in a short period of time.

Iodine Fortification

Iodine deficiency is present in almost all parts of the developed and developing world, and environmental iodine deficiency is the main cause of iodine deficiency disorders. Iodine is irregularly distributed over the earth's crust, resulting in acute deficiencies in areas such as mountainous regions and flood plains. The problem is aggravated by accelerated deforestation and soil erosion. Thus, the food grown in iodine-deficient regions can never provide enough iodine for the people and livestock living there. The iodine deficiency results from geologic rather than social and economic conditions. It cannot be eliminated by changing dietary habits or by eating specific kinds of foods but must be corrected by supplying iodine from external sources. It has, therefore, been a common practice to use common salt as a vehicle for iodine fortification for the past 75 years. Salt is consumed at approximately the same level throughout the year by the entire population of a region. Universal salt iodisation is now a widely accepted strategy for preventing and correcting iodine deficiency disorders.

There are areas where consumption of goitrogens in the staple diet (e. g., cassava) affects the proper utilisation of iodine by the thyroid gland. For example, in Congo, Africa, as a result of cassava diets there is an overload of thiocyanate. To overcome this problem, appropriate increases in salt iodisation are required to ensure the recommended dietary intake. The iodisation of salt is done either by spraying potassium iodate or potassium iodide in amounts that ensure a minimum of 150 μg iodine/day. Both of these forms of iodine are absorbed as iodide ions and are completely bio-available. Other methods of iodine prophylaxis are also used: iodised oil (capsule and injections), iodised water, iodised bread, iodised soya sauce, iodoform compounds used in dairy and poultry, and certain food additives.

Iodine loss occurs as a result of improper packaging, Humidity and moisture, and transport in open trucks and railway wagons exposed to sunlight. To compensate for these losses, higher levels of iodine are used during the production of iodised salt. Losses during the cooking process vary from 20 percent to 40 percent depending on the type of cooking used.

To ensure the consumption of recommended levels of iodine, the iodine content of salt at the production level should be monitored with proper quality assurance programmes. Regular evaluation of the urinary iodine excretion pattern in the population consuming iodised salt or exposed to other iodine prophylactic measures would help the adjusting of iodine intake.

IRON

The Role of Iron in Human Metabolic Processes

Iron has several vital functions in the body. It serves as a carrier of oxygen to the tissues from the lungs by red blood cell haemoglobin, as a transport medium for electrons within cells, and as an integrated part of important enzyme systems in various tissues. The physiology of iron has been extensively reviewed (1-6).

Most of the iron in the body is present in the erythrocytes as haemoglobin, a molecule composed of four units, each containing one heme group and one protein chain. The structure of haemoglobin allows it to be fully loaded with oxygen in the lungs and partially unloaded in the tissues (e. g., in the muscles). The iron-containing oxygen storage protein in the muscles, myoglobin, is similar in structure to haemoglobin but has only one heme unit and one globin chain. Several iron-containing enzymes, the cytochromes, also have one heme group and one globin protein chain.

These enzymes act as electron carriers within the cell and their structures do not permit reversible loading and unloading of oxygen. Their role in the oxidative metabolism is to transfer energy within the cell and specifically in the mitochondria. Other key functions for the iron-containing enzymes (e. g., cytochrome P450) include the synthesis of steroid hormones and bile acids; detoxification of foreign substances in the liver; and signal controlling in some neurotransmitters, such as the dopamine and serotonin systems in the brain. Iron is reversibly stored within the liver as ferritin and hemosiderin whereas it is transported between different compartments in the body by the protein transferrin.

Iron Requirements

Basal Iron Losses

Iron is not actively excreted from the body in urine or in the intestines. Iron is only lost with cells from the skin and the interior surfaces of the body - intestines, urinary tract, and airways.

The total amount lost is estimated at 14 μg/kg body weight/day. In children, it is probably more correct to relate these losses to body surface. A non-menstruating 55-kg women loses about 0. 8 mg Fe/day and a 70-kg man loses about 1 mg. The range of individual variation has been estimated to be ±15 percent.

Earlier studies suggested that sweat iron losses could be considerable, especially in a hot, Humid climate. However, new studies which took extensive precautions to avoid the interference of contamination of iron from the skin during the collection of total body sweat have shown that these sweat iron losses are negligible.

Growth

The newborn term infant has an iron content of about 250-300 mg (75 mg/kg body weight). During the first 2 months of life, haemoglobin concentration falls because of the improved oxygen situation in the newborn infant compared with the intrauterine foetus. This leads to a considerable redistribution of iron from catabolised erythrocytes to iron stores.

This iron will cover the needs of the term infant during the first 4-6 months of life and is why iron requirements during this period can be provided by human milk, that contains very little iron. Because of the marked supply of iron to the foetus during the last trimester of pregnancy, the iron situation is much less favourable in the premature and low-birth-weight infant than in the term infant. An extra supply of iron is therefore needed in these infants even during the first 6 months of life.

In the full-term infant, iron requirements will rise markedly after age 4-6 months and amount to about 0. 7-0. 9 mg/day during the remaining part of the first year. These requirements are therefore very high, especially in relation to body sise and energy intake.

Table: Iron intakes required for growth under the age of 18 years, median basal iron losses, menstrual losses in women, and total absolute iron requirements

					Menstrual losses		Total Absolute Requirements	
Group	Age	Body weighti Mean Growth	Required ntakes for Median	Basal Iron losses	Median	95th per-centile	Median	95th per-centile
	(Ye-)	(kg)	(mg/ day)	(mg/ day)	(mg/ day)	(mg/ day)	(mg/ day)	(mg/ day)
Chil-dren	0. 5-1	9	0. 55	0. 17	0	0	0. 72	0. 93
	1-3	13. 3	0. 27	0. 19	0	0	0. 46	0. 58
	4-6	19. 2	0. 23	0. 27	0	0	0. 50	0. 63
	7-10	28. 1	0. 32	0. 39	0	0	0. 71	0. 89
Males	11-14	45	0. 55	0. 62	0	0	1. 17	1. 46
	15-17	64. 4	0. 60	0. 90	0	0	1. 50	1. 88
	18+	75	0	1. 05	0	0	1. 05	1. 37
Fem ales	11-14	[b]46. 1	0. 55	0. 65	0	0	1. 20	1. 40
	11-14	46. 1	0. 55	0. 65	0. 48[c]	1. 90[c]	1. 68	3. 27
	15-17	56. 4	0. 35	0. 79	0. 48[c]	1. 90[c]	1. 62	3. 10
	18+	62	0. 87	0. 48[c]	1. 90[c]	1. 46	2. 94	
Post-meno pausal	0	62	0	0. 87	0	0	0. 87	1. 13
Lact ating	0	62	0	1. 15	0	0	1. 15	1. 50

Total Absolute Requirements = Requirement for growth + basal losses + menstrual losses (females only)

[a]Based in part on a 1988 report from Food and Agriculture Organization of the United Nations (FAO)/World Health Organizaiton (WHO) and in part on new calculations of the distribution of iron requirements in menstruating women.

Because of the very skewed distribution of iron requirements in these women, dietary iron requirements are calculated for four levels of dietary iron bio-availability.

[b]Non-menstruating.

[c]Effect of the normal variation in haemoglobin concentration not included in this figure.

In the first year of life, the full-term infant almost doubles its total iron stores and triple its body weight. The change in body iron during this period occurs mainly during the first 6-12 months of life. Between 1 and 6 years of age, the body iron content is again doubled. The requirements for absorbed iron in infants and children are very high in relation to their energy requirements. For example, in infants 6-12 months of age, about 1. 5 mg of iron need to be absorbed per 4. 184 MJ and about half of this amount is required up to age 4 years.

Table: The recommended nutrient intakes for iron based on varying dietary iron bio-availabilities

Group	Age (years)	Mean Body weight (kg)	Recommended Nutrient Intake[a](mg/day) % Dietary Iron Bio-vailability 15	12	10	5
Children	0. 5-1	9	[6. 2][b]	[7. 7][b]	[9. 3][b]	[18. 6][b]
	1-3	13. 3	3. 9	4. 8	5. 8	11. 6
	4-6	19. 2	4. 2	5. 3	6. 3	12. 6
	7-10	28. 1	5. 9	7. 4	8. 9	17. 8
Males	11-14	45	9. 7	12. 2	14. 6	29. 2
	15-17	64. 4	12. 5	15. 7	18. 8	37. 6
	18+	75	9. 1	11. 4	13. 7	27. 4
Females	11-14[c]	46. 1	9. 3	11. 7	14	28
	11-14	46. 1	21. 8	27. 7	32. 7	65. 4
	15-17	56. 4	20. 7	25. 8	31	62
	18+	62	19. 6	24. 5	29. 4	58. 8
Post-men opausal	0	62	7. 5	9. 4	11. 3	22. 6
Lactating	0	62	10	12. 5	15	30

- Based in part on a 1988 report from the FAO/WHO and in part on new calculations of the distribution of iron requirements in menstruating women. Because of the very skewed distribution of iron requirements in these women, dietary iron requirements are calculated for four levels of dietary iron bio-availability.
- Bio-availability of dietary iron during this period varies greatly.
- Non-menstruating.

In the weaning period, the iron requirements in relation to energy intake are the highest of the lifespan except for the last trimester of pregnancy, when iron requirements to a large extent have to be covered from the iron stores of the mother. The rapidly growing weaning infant has no iron stores and has to rely on dietary iron. It is possible to meet these high requirements if the diet has a consistently high content of meat and foods rich in ascorbic acid. In most developed countries today, infant cereal products are the staple foods for that period of life. Commercial products are regularly fortified with iron and ascorbic acid, and they are usually given together with fruit juices and solid foods containing meat, fish, and vegetables. The fortification of cereal products with iron and ascorbic acid is important in meeting the high dietary needs, especially considering the importance of an optimal iron nutrition during this phase of brain development.

Iron requirements are also very high in adolescents, particularly during the period of rapid growth. There is a marked individual variation in growth rate and the requirements may be considerably higher than the calculated mean values given in Table. Girls usually have their growth spurt before menarche, but growth is not finished at that time. Their total iron requirements are therefore considerable. In boys during puberty there is a marked increase in haemoglobin mass and concentration, further increasing iron requirements to a level above the average iron requirements in menstruating women.

Menstrual Iron Losses

Menstrual blood losses are very constant from month to month for an individual but vary markedly from one woman to another. The main part of this variation is genetically controlled by the content of fibrinolytic activators in the uterine mucosa even in populations which are geographically widely separated.

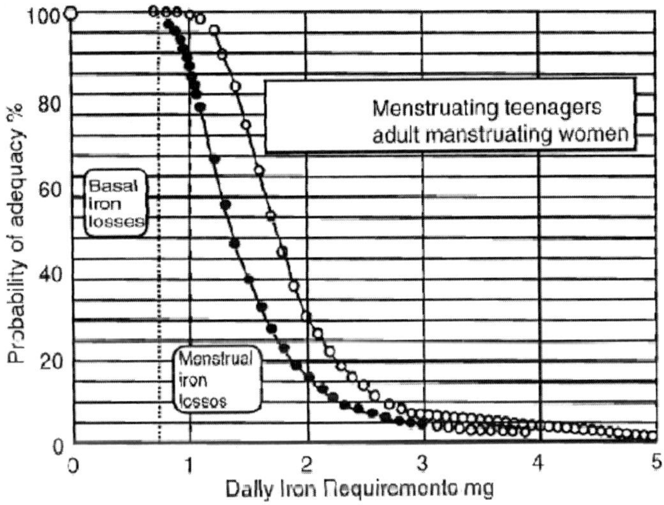

Fig. Distribution of daily iron requirements in menstruating adult women and teenagers: the probability of adequacy at different amounts of iron absorbed

These findings strongly suggest that the main source of variation in iron status in different populations is not related to a variation in iron requirements but to a variation in the absorption of iron from the diets. (This statement disregards infestations with hookworms and other parasites). The mean menstrual iron loss, averaged over the entire menstrual cycle of 28 days, is about 0. 56 mg/day. The frequency distribution of physiologic menstrual blood losses is highly skewed. Adding the average basal iron loss (0. 8 mg) and its variation allows the distribution of the total iron requirements in adult women to be calculated as the convolution of the distributions of menstrual and basal iron losses.

The mean daily total iron requirement is 1. 36 mg. In 10 percent of women it exceeds 2. 27 mg and in 5 percent it exceeds 2. 84 mg. In 10 percent of menstruating (still-growing) teenagers, the corresponding daily total iron requirement exceeds 2. 65 mg, and in 5 percent of the girls it exceeds 3. 2 mg/day. The marked skewness of menstrual losses is a great nutritional problem because personal assessment of the losses is unreliable.

This means that women with physiologic but heavy losses cannot be identified and reached by iron supplementation. The choice of contraceptive method greatly influences menstrual losses. The methods of calculating iron requirements in women and their variation were recently re-examined.

In postmenopausal women and in physically active elderly people, the iron requirements per unit of body weight are the same as in men. When physical activity decreases as a result of ageing, blood volume and haemoglobin mass also diminish, leading to a shift of iron from haemoglobin and muscle to iron stores.

This implies a reduction of the daily iron requirements. Iron deficiency in the elderly is therefore seldom of nutritional origin but is usually caused by pathologic iron losses. The absorbed iron requirements in different groups are given in Table. The iron situation during pregnancy and lactation are dealt with separately below.

Iron Absorption

With respect to the mechanism of absorption, there are two kinds of dietary iron: heme iron and non-heme iron. In the human diet the primary sources of heme iron are the haemoglobin and myoglobin from consumption of meat, poultry, and fish whereas non-heme iron is obtained from cereals, pulses, legumes, fruits, and vegetables. The average absorption of heme iron from meat-containing meals is about 25 percent The absorption of heme iron can vary from about

40 percent during iron deficiency to about 10 percent during iron repletion. Heme iron can be degraded and converted to non-heme iron if foods are cooked at a high temperature for too long. Calcium is the only dietary factor that negatively influences the absorption of heme iron and does so to the same extent that it influences non-heme iron.

Enhancing factors Ascorbic acid (e. g., certain fruit juices, fruits, potatoes, and certain vegetables)Meat, chicken, fish and other seafood Fermented vegetables (e. g., sauerkraut), fermented soy sauces, etc. Inhibiting factors Phytates and other inositol phosphates (e. g., bran products, bread made from high-extraction flour, breakfast cereals, oats, rice [especially unpolished rice], pasta products, cocoa, nuts, soya beans, and peas) Iron-binding phenolic compounds (e. g., tea, coffee, cocoa, certain spices, certain vegetables, and most red wines) Calcium (e. g., milk, cheese) Soy proteins

Non-heme iron is the main form of dietary iron. The absorption of non-heme iron is influenced by individual iron status and by several factors in the diet. Dietary factors influencing iron absorption are outlined in Table. Iron compounds used for the fortification of foods will only be partially available for absorption.

Once iron is dissolved, its absorption from fortificants and food contaminants is influenced by the same factors as the iron native to the food substance. Iron originating from the soil (e. g., from various forms of clay) is sometimes present in considerable amounts on the surface of foods as a contaminant originating from dust on air-dried foods or from water used in irrigation. Even if the fraction of iron that is available is often small, contamination iron may still be nutritionally important because of the great amounts present.

Reducing substances (i. e., substances that keep iron in the ferrous form) must be present for iron to be absorbed. The presence of meat, poultry, and fish in the diet enhance iron absorption. Other foods contain factors (ligands) that strongly bind ferrous ions, that subsequently inhibit absorption. Examples are phytates and certain iron-binding polyphenols.

Inhibition of Iron Absorption

Phytates are found in all kinds of grains, seeds, nuts, vegetables, roots (e. g., potatoes), and fruits. Chemically, phytates are inositol hexaphosphate salts and are a storage form of phosphates and minerals. Other phosphates have not been shown to inhibit non-heme iron absorption. In North American and European diets, about 90 percent of phytates originate from cereals. Phytates strongly inhibit iron absorption in a dose-dependent fashion and even small amounts of phytates have a marked effect.

Bran has a high content of phytate and strongly inhibits iron absorption. Whole-wheat flour, therefore, has a much higher content of phytates than does white wheat flour. In bread some of the phytates in bran are degraded during the fermentation of the dough. Fermentation for a couple of days (sourdough fermentation) can therefore almost completely degrade the phytate and increase the bio-availability of iron in bread made from whole-wheat flour. Oats strongly inhibit iron absorption because of their high phytate content, that results from native phytase in oats being destroyed by the normal heat process used to avoid rancidity. Sufficient amounts of ascorbic acid can counteract this inhibition. By contrast, non-phytate-containing dietary fibre components have almost no influence on iron absorption.

Almost all plants contain phenolic compounds as part of their defence system against insects, animals, and humans. Only some of the phenolic compounds (mainly those containing galloyl groups) seem to be responsible for the inhibition of iron absorption. Tea, coffee, and cocoa are common plant products that contain iron-binding polyphenols. Many vegetables, especially green leafy vegetables (e. g., spinach), and herbs and spices (e. g., oregano) contain appreciable amounts of galloyl groups, that strongly inhibit iron absorption. Consumption of betel leaves, common in areas of Asia, also has a marked negative effect on iron absorption.

Calcium, consumed as a salt or in dairy products interferes significantly with the absorption of both heme and non-heme iron. Because calcium and iron are both essential nutrients, calcium cannot be considered to be an inhibitor in the same way as phytates or phenolic compounds. The practical solution for this competition is to increase iron intake, increase its bio-availability, or avoid the intake of foods rich in calcium and foods rich in iron at the same meal.

The mechanism of action for absorption inhibition is unknown, but the balance of evidence strongly suggest that the inhibition is located within the mucosal cell itself at the common final transfer step for heme and non-heme iron. Recent analyses of the dose-effect relationship show that no inhibition is seen from the first 40 mg of calcium in a meal. A sigmoid relationship is then seen, reaching a 60 percent maximal inhibition of iron absorption by 300-600 mg calcium. The form of this curve suggests a one-site competitive binding of iron and calcium. This relationship explains some of the seemingly conflicting results obtained in studies on the interaction between calcium and iron.

For unknown reasons, the addition of soy protein to a meal reduces the fraction of iron absorbed. This inhibition is not solely explained by the high phytate content of soy protein. However, because of the high iron content of soy proteins, the net effect on iron absorption of an addition of soy products to a meal is usually positive. In infant foods containing soy proteins, the inhibiting effect can be overcome by the addition of sufficient amounts of ascorbic acid. Some fermented soy sauces, however, have been found to enhance iron absorption.

Enhancement of Iron Absorption

Ascorbic acid is the most potent enhancer of non-heme iron absorption. Synthetic vitamin C increases the absorption of iron to the same extent as the native ascorbic acid in fruits, vegetables, and juices. The effect of ascorbic acid on iron absorption is so marked and essential that this effect could be considered as one of vitamin C's physiologic roles. Each meal should preferably contain at least 25 mg of ascorbic acid and possibly more if the meal contains many inhibitors of iron absorption. Therefore, a requirement of ascorbic acid for iron absorption should be taken into account when establishing the requirements for vitamin C, that are set only to prevent vitamin C deficiency.

Meat, fish, and seafood all promote the absorption of non-heme iron. The mechanism for this effect has not been determined. It should be pointed out that meat also enhances the absorption of heme iron to about the same extent. Meat promotes iron nutrition in two ways: it stimulates the absorption of both heme and non-heme iron and it provides the well-absorbed heme iron. Epidemiologically, the intake of meat has been found to be associated with a lower prevalence of iron deficiency.

Organic acids, such as citric acid, have in some studies been found to enhance the absorption of non-heme iron. This effect is not observed as consistently as is the effect of ascorbic acid.

Iron Absorption from Meals

The pool concept in iron absorption implies that there are two main pools in the gastrointestinal lumen - one pool of heme iron and another pool of non-heme iron - and that iron absorption takes place independently from these two pools. The pool concept also implies that the absorption of iron from the non-heme iron pool results from all ligands present in the mixture of foods included in a meal.

The absorption of non-heme iron from a certain meal not only depends on its iron content but also, and to a marked degree, on the composition of the meal (i. e., the balance among all factors enhancing and inhibiting the absorption of iron). The bio-availability can vary more than 10-fold among meals with a similar content of iron, energy, protein, fat, etc. Just the addition of certain spices (e. g., oregano) or a cup of tea may reduce the bio-availability by one-half or more. However, the addition of certain vegetables or fruits containing ascorbic acid may double or even triple iron absorption, depending on the other properties of the meal and the amounts of ascorbic acid present.

Iron Absorption from the Whole Diet

There is limited information about the total amounts of iron absorbed from the diet because no simple method is available to measure iron absorption from the whole diet. It has been measured by chemical balance studies using long balance periods or by determining the haemoglobin regeneration rate in subjects with induced iron deficiency anaemia and a well-controlled diet over a long period of time.

A method was recently developed to measure iron absorption from the whole diet. In the first studies all non-heme iron in all meals over periods of 5-10 days was homogeneously labelled to the same specific activity with an extrinsic inorganic radioiron tracer. Heme iron absorption was then estimated. In a further study, heme and non-heme iron were separately labelled with two radioiron tracers as biosynthetically labelled haemoglobin and as an inorganic iron salt. New information could be obtained, for example, about the average bio-availability of dietary iron in different types of diets, overall effects of certain factors (e. g., calcium) on iron nutrition, and regulation of iron absorption in relation to iron status. Iron absorption from the whole diet is the sum of the absorption of iron from the single meals included in the diet.

It has been suggested that the iron absorption of single meals may exaggerate the absorption of iron from the diet. Iron absorption from single meals can never represent iron absorption from the whole diet, but iron absorption from a single meal was the same when the meal was served in the morning after an overnight fast or at lunch or supper. The same observation was made in another study when a hamburger meal was served in the morning or 2-4 hours after a breakfast.

Because energy expenditure and energy intake set the limit for the amount of food eaten and for meal sise, it is practical to relate the bio-availability of iron in different meals to energy content (bio-available nutrient density). The use of bio-available nutrient density is a feasible way to compare different meals, construct menus, and calculate recommended intakes.

Intake of energy and essential nutrients such as iron was probably considerably higher for early humans than it is today. The present low iron intake associated with a low-energy lifestyle implies that the interaction between different factors influencing iron absorption, will be more critical. For example, the interaction between calcium and iron absorption probably had no importance in the nutrition of early humans, who had a diet with ample amounts of both iron and calcium.

Iron Balance and Regulation of Iron Absorption

The body has three unique mechanisms for maintaining iron balance and preventing iron deficiency and iron overload. The first is the continuous re-utilisation of iron from catabolised erythrocytes in the body. When an erythrocyte dies after about 120 days, it is usually degraded by the macrophages of the reticular endothelium. The iron is released and delivered to transferrin in the plasma, which brings the iron back to red blood cell precursors in the bone marrow or to other cells in different tissues. Uptake and distribution of iron in the body is regulated by the synthesis of transferrin receptors on the cell surface. This system for internal iron transport not only controls the rate of flow of iron to different tissues according to their needs but also effectively prevents the appearance of free iron and the formation of free radicals in the circulation.

The second mechanism is the access of the specific storage protein, ferritin, which can store and release iron to meet excessive iron demands. This iron reservoir is especially important in the third trimester of pregnancy.

The third mechanism involves the regulation of absorption of iron from the intestines, with an increased iron absorption in the presence of decreasing body iron stores and a decreased iron absorption when iron stores increase. Iron absorption decreases until an equilibrium is established between absorption and requirements.

For a given diet this regulation of iron absorption, however, can only balance losses up to a certain critical point beyond which iron deficiency will develop. About half of the basal iron losses are from blood, primarily in the gastrointestinal tract. Both these losses and the menstrual iron losses are influenced by the haemoglobin level; during the development of an iron deficiency, menstrual and basal iron losses will successively decrease when the haemoglobin level decreases. In a state of more severe iron deficiency, skin iron losses may also decrease. Iron balance (absorption equals losses) may be present not only in normal subjects but also during iron deficiency and iron overload.

The three main factors that affect iron balance are absorption (intake and bio-availability of iron), losses, and amount in stores. The interrelationship among these factors was recently been described in mathematical terms, making it possible to predict, for example, the amount of stored iron when iron losses and bio-availability of dietary iron are known. With increasing iron requirements or decreasing bio-availability, the regulatory capacity to prevent iron deficiency is limited. However, to prevent iron overload with increasing dietary iron intake or bio-availability, the regulatory capacity seems to be extremely good.

Iron Deficiency

Populations at Risk for Iron Deficiency

Worldwide, the highest prevalence of iron deficiency is found in infants, children, adolescents, and women of childbearing age, especially pregnant women. The weaning period in infants is especially critical because of the very high iron requirements in relation to energy requirements. Thanks to better information and access to fortified cereals for infants and children, the iron situation has markedly improved in these groups in most industrialised countries where the highest prevalences of iron deficiency today are observed in menstruating and pregnant women and adolescents of both sexes. In developing countries, however, the iron situation is very critical in many groups, especially in the weaning period. Iron nutrition is of great importance for the adequate development of the brain and other tissues such as muscles, which are finally differentiated early in life.

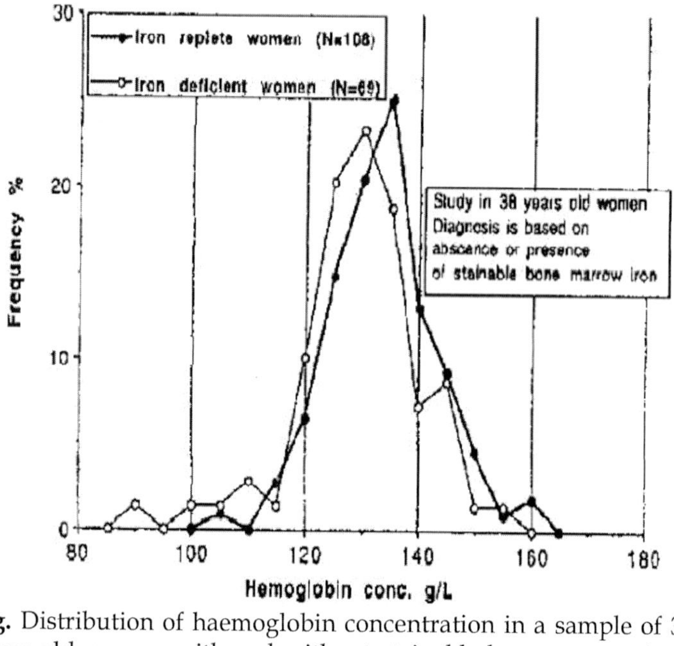

Fig. Distribution of haemoglobin concentration in a sample of 38-year-old women with and without stainable bone marrow iron

Iron deficiency and iron deficiency anaemia are often incorrectly used as synonyms. A definition of these terms may clarify some confusion about different prevalence figures given in the literature. The cause of the problem is the very wide distribution of the haemoglobin concentration in healthy, fully iron-replete subjects. During the development of a negative iron balance in subjects with no mobilisable iron from iron stores (no visible iron in technically perfect bone marrow smears or a serum ferritin concentration <15 μg/l), there will be an immediate impairment in the production of haemoglobin with a resulting decrease in haemoglobin and different erythrocyte indexes (e. g., mean corpuscular haemoglobin and mean corpuscular volume). In turn this will lead to an overlap of the distributions of haemoglobin in iron-deficient and iron-replete women. The extent of overlap depends on the prevalence and severity of iron deficiency. In populations with more severe iron deficiency, for example, the overlap is much less marked.

In women, anaemia is defined as a haemoglobin <120 g/l. For a woman who has her normal homeostatic value set at 150 g/l, haemoglobin level must decrease to 119 g/l (by 26 percent) before she is considered to be anaemic, whereas for a woman who has her normal haemoglobin set at 121 g/l, haemoglobin level must only decrease by 1. 5 percent to 119 g/l. Iron deficiency anaemia is a rather imprecise concept for evaluating the single subject and has no immediate physiologic meaning. By definition, this implies that the prevalence of iron deficiency anaemia is less frequent than iron deficiency and that the presence of anaemia in a subject is a statistical rather than a functional concept.

The main use of the cut-off value is in comparisons between population groups. In practical work, iron deficiency anaemia should be replaced by the functional concept of iron deficiency. Anaemia per se is mainly important when it becomes so severe that oxygen delivery to tissues is impaired. An iron deficiency anaemia which develops slowly in otherwise healthy subjects with moderately heavy work output will not give any symptoms until the haemoglobin level is about 80 g/l or lower. The reason for the continued use of the concept of iron deficiency anaemia is the ease of determining haemoglobin. Therefore, in clinical practice, knowledge of previous haemoglobin values in a subject is of great importance for evaluating the diagnosis.

Iron deficiency being defined as an absence of iron stores combined with signs of an iron-deficient erythropoiesis implies that in a state of iron deficiency there is an insufficient supply of iron to various tissues. This occurs at a serum ferritin level <15 μg/l. Iron can then no longer be mobilised from iron stores and insufficient amounts of iron will be delivered to transferrin, the circulating transport protein for iron.

The binding sites for iron on transferrin will therefore contain less and less iron. This is usually described as a reduction in transferrin saturation. When transferrin saturation drops to a certain critical level, erythrocyte precursors, which continuously need iron for the formation of haemoglobin, will get an insufficient supply of iron.

At the same time, the supply of iron by transferrin to other tissues will also be impaired. Liver cells will get less iron, more transferrin will be synthesised, and the concentration of transferrin in plasma will then suddenly increase. Cells with a high turnover rate are the first ones to be affected (e. g., intestinal mucosal cells with a short life span).

The iron-transferrin complex is bound to transferrin receptors on cell surfaces and the whole complex is then taken up by special receptors on the surface of various cells and tissues. The uptake of iron seems to be related both to transferrin saturation and the number of transferrin receptors on the cell surface. There is a marked diurnal variation in the saturation of transferrin because the turnover rate of iron in plasma is very high. This fact makes it difficult to evaluate the iron status from single determinations of transferrin saturation.

Indicators of Iron Deficiency

The absence of iron stores (iron deficiency) can be diagnosed by showing that there is no stainable iron in the reticuloendothelial cells in bone marrow smears or more easily by a low concentration of ferritin in serum (£15 μg/l). Even if an absence of iron stores per se may not necessarily be associated with any immediate adverse effects, it is a reliable and good indirect indicator of iron-deficient erythropoiesis and of an increased risk of a compromised supply of iron to different tissues.

Even before iron stores are completely exhausted, the supply of iron to the erythrocyte precursors in the bone marrow is compromised, leading to iron-deficient erythropoiesis. A possible explanation is that the rate of release of iron from stores is influenced by the amount of iron remaining. As mentioned above it can then be assumed that the supply of iron to other tissues needing iron is also insufficient because the identical transport system is used. During the development of iron deficiency haemoglobin concentration, transferrin concentration, transferrin saturation, transferrin receptors in plasma, erythrocyte protoporphyrin, and erythrocyte indexes are changed.

All these methods, however, show a marked overlap between normal and iron-deficient subjects, that makes it impossible to identify the single subject with mild iron deficiency by using any of these methods. Therefore, these tests have been used in combination (e. g., for interpreting results from the second National Health and Nutrition Examination Survey in the United States of America [75, 76]). The diagnostic specificity then increases but the sensitivity decreases, and thus the true prevalence of iron deficiency is markedly underestimated if multiple diagnostic criteria are used. By definition in screening for iron deficiency, the more tests that are used the higher is the diagnostic specificity but the lower is the sensitivity of the procedure. Fortunately, a low serum ferritin, £15μg/l is always associated with an iron-deficient erythropoiesis. The use of serum ferritin alone as a measure will also underestimate the true prevalence of iron deficiency but to a lesser degree than when the combined criteria are used.

A diagnosis of iron deficiency anaemia can be suspected if anaemia is present in subjects who are iron-deficient as described above. Preferably, to fully establish the diagnosis, the subjects should respond adequately to iron treatment. The pitfalls with this method are the random variation in haemoglobin concentrations over time and the effect of the regression towards the mean when a new measurement is made.

The use of serum ferritin has improved the diagnostic accuracy of iron deficiency. It is the only simple method available to detect early iron deficiency. Its practical value is somewhat reduced, however, by the fact that serum ferritin is a very sensitive acute-phase reactant and may be increased for weeks after a simple infection with fever for a day or two. Several other conditions, such as use of alcohol, liver disease, and collagen diseases, may also increase serum ferritin concentrations.

Determination of transferrin receptors in plasma has also been recommended in the diagnosis of iron deficiency. Its advantage is that it is not influenced by infections. Its main use is in subjects who are already anaemic and it is not sensitive for the early diagnosis of iron deficiency. The use of a combination of determinations of serum ferritin and serum transferrin receptors has also been suggested.

Causes of Iron Deficiency

Nutritional iron deficiency implies that the diet cannot cover physiologic iron requirements. Worldwide this is the most common cause of iron deficiency. In many tropical countries, infestations with hookworms lead to intestinal blood losses that may be considerable. The severity of the infestations with hookworms varies considerably between subjects and regions. The average blood loss can be well estimated by egg counts in stools. Usually the diet in these populations is also limited with respect to iron content and availability.

In clinical practice a diagnosis of iron deficiency must always lead to a search for pathologic causes of blood loss (e. g., tumours in the gastrointestinal tract or uterus, especially if uterine bleedings have increased or changed in regularity). Patients with achlorhydria absorb dietary iron less well (a reduction of about 50 percent) and patients who have undergone gastric surgery, especially if the surgery was extensive, may eventually develop iron deficiency because of impaired iron absorption. Gluten enteropathy is another possibility to consider, especially in young patients.

Prevalence of Iron Deficiency

Iron deficiency is probably the most frequent nutritional deficiency disorder in the world. A recent estimate based on World Health Organization (WHO) criteria indicated that around 600-700 million people worldwide have a marked iron deficiency anaemia. In industrialised countries, the prevalence of iron deficiency anaemia is much lower and usually varies between 2 percent and 8 percent. However, the prevalence of iron deficiency, including both anaemic and non-anaemic subjects, is much higher. In industrialised countries, for example, an absence of iron stores or subnormal serum ferritin values is found in about 20-30 percent of women of fertile age. In adolescent girls the prevalence is even higher.

It is difficult to determine the prevalence of iron deficiency more exactly because representative populations for clinical investigation are hard to obtain. Labouratory methods and techniques for blood sampling need careful standardization. One often neglected source of error, for example, when materials in different regions or at different times are compared, is the fact that there are still reagent kits on the market for determining serum ferritin which are not adequately calibrated at different concentrations against the international WHO standards. In addition, seasonal variations in infection rates influence the sensitivity and specificity of most methods used.

Worldwide, the highest prevalence figures for iron deficiency are found in infants, children, teenagers, and women of childbearing age. Thanks to better information and access to fortified cereals for infants and children, the iron situation has markedly improved in these groups in most industrialised countries, where the highest prevalence today is observed in menstruating women and adolescents of both sexes.

In developing countries, where the prevalence of iron deficiency is very high and the severity of anaemia is marked, studies on the distribution of haemoglobin in different population groups can provide important information as a valuable basis for action programmes. A more detailed analysis of subsamples may then give excellent information for the planning of more extensive programmes.

Effects of Iron Deficiency

Studies in animals have clearly shown that iron deficiency has several negative effects on important functions in the body. Physical working capacity in rats has been shown to be significantly reduced in iron deficiency, that is especially valid for endurance activities. This negative effect seems to be less related to the degree of anaemia than to impaired oxidative metabolism in the muscles with an increased formation of lactic acid, that in turn is due to a lack of iron-containing enzymes which are rate limiting for the oxidative metabolism.

The relationship between iron deficiency and brain function is of great importance for the choice of strategy in combating iron deficiency. Several structures in the brain have a high iron content of the same magnitude as observed in the liver. Of great importance is the observation that the lower iron content of the brain in iron-deficient growing rats cannot be increased by giving iron later on. This fact strongly suggests that the supply of iron to brain cells takes place during an early phase of brain development and that, as such, early iron deficiency may lead to irreparable damage to brain cells.

In humans about 10 percent of brain iron is present at birth; at the age of 10 years the brain has only reached half its normal iron content, and optimal amounts are first reached at the age of 20-30 years.

In populations with long-standing iron deficiency, a reduction of physical working capacity has been demonstrated by several groups with improvement in working capacity after iron administration.

Iron deficiency also negatively influences the normal defence systems against infections. The cell-mediated immunologic response by the action of T lymphocytes is impaired as a result of a reduced formation of these cells. This in turn is due to a reduced DNA synthesis depending on the function of ribonucleotide reductase, which requires a continuous supply of iron for its function. The phagocytosis and killing of bacteria by the neutrophil leukocytes is an important component of the defence mechanism against infections. These functions are impaired in iron deficiency. The killing function is based on the formation of free hydroxyl radicals within the leukocytes, the respiratory burst, and results from the activation of the iron-sulphur enzyme NADPH oxidase and probably also cytochrome b (a heme enzyme).

The impairment of the immunologic defence against infections that was found in animals is also regularly found in humans. Administration of iron normalises these changes within 4-7 days. It has been difficult to demonstrate, however, that the prevalence of infections is higher or that their severity is more marked in iron-deficient subjects than in control subjects. This may well be ascribed to the difficulty in studying this problem with an adequate experimental design.

A relationship between iron deficiency and behaviour such as attention, memory, and learning, has been demonstrated in infants and small children by several groups. In the most recent well-controlled studies, no effect was noted from the administration of iron. This finding is consistent with the observations in animals. Therapy-resistant behavioural impairment and the fact that there is an accumulation of iron during the whole period of brain growth should be considered strong arguments for the more active and effective combating of iron deficiency. This is valid for women, especially during pregnancy, for infants and children, and up through the period of adolescence and early adulthood. In a recent well-controlled study, administration of iron to non-anaemic but iron-deficient adolescent girls improved verbal learning and memory.

Well-controlled studies in adolescent girls show that iron-deficiency without anaemia is associated with reduced physical endurance and changes in mood and ability to concentrate. A recent careful study showed that there was a reduction in maximum oxygen consumption in non-anaemic women with iron deficiency that was unrelated to a decreased oxygen-transport capacity of the blood.

Iron During Pregnancy and Lactation

Iron requirements during pregnancy are well established. Most of the iron required during pregnancy is used to increase the haemoglobin mass of the mother, which occurs in all healthy pregnant women who have sufficiently large iron stores or who are adequately supplemented with iron. The increased haemoglobin mass is directly proportional to the increased need for oxygen transport during pregnancy and is one of the important physiologic adaptations that occurs in pregnancy. A major problem for iron balance in pregnancy is that iron requirements are not equally distributed over its duration. The exponential growth of the foetus implies that iron needs are almost negligible in the first trimester and that more than 80 percent relates to the last trimester. The total daily iron requirements, including the basal iron losses (0. 8 mg), increase during pregnancy from 0. 8 mg to about 10 mg during the last 6 weeks of pregnancy.

Iron absorption during pregnancy is determined by the amount of iron in the diet, its bio-availability (meal composition), and the changes in iron absorption that occur during pregnancy. There are marked changes in the fraction of iron absorbed during pregnancy. In the first trimester there is a marked, somewhat paradoxical, decrease in the absorption of iron, which is closely related to the reduction in iron requirements during this period as compared with the non-pregnant state. In the second trimester iron absorption is increased by about 50 percent, and in the last trimester it may increase by up to about four times. Even considering the marked increase in iron absorption, it is impossible for the mother to cover her iron requirements from diet alone, even if its iron content and bio-availability are very high. It can be calculated that with diets prevailing in most industrialised countries, there will be a deficit of about 400-500 mg in the amount of iron absorbed during pregnancy.

An adequate iron balance can be achieved if iron stores of 500 mg are available. However, it is uncommon for women today to have iron stores of this sise. It is therefore recommended that iron supplements in tablet form, preferably together with folic acid, be given to all pregnant women because of the difficulties in correctly evaluating iron status in pregnancy with routine labouratory methods. In the non-anaemic pregnant woman, daily supplements of 100 mg of iron (e. g., as ferrous sulphate) given during the second half of pregnancy are adequate. In anaemic women higher doses are usually required.

Table: Iron requirements during pregnancy

Iron requirements (mg)	
IRON REQUIREMENTS DURING PREGNANCY	300
Foetus Placenta	50
Expansion of maternal erythrocyte mass	450
Basal iron losses	240
Total iron requirement	1040
NET IRON BALANCE AFTER DELIVERY	
Contraction of maternal erythrocyte mass	+450
Maternal blood loss	-250
Net iron balance	+200
Net iron requirements for pregnancy if sufficient maternal iron stores are present (1040 - 200 = 840)	840

During the birth process, the average blood loss corresponds to about 250 mg iron. At the same time, however, the haemoglobin mass of the mother is gradually normalised, which implies that about 200 mg iron from the expanded haemoglobin mass (150-250 mg) is returned to the mother. To cover the needs of a woman after pregnancy, a further 300 mg of iron must be accumulated in the iron stores in order for the woman to start her next pregnancy with about 500 mg of stored iron. Such a restitution is not possible with present types of diets.

There is an association between low haemoglobin values and prematurity. An extensive study showed that a woman with a hematocrit of 37 percent had twice the risk of having a premature birth as did a woman with a hematocrit between 41 percent and 44 percent (P£0. 01). A similar observation was reported in another extensive study. These materials were examined retrospectively and the cause of the lower hematocrit was not examined.

In lactating women, the daily iron loss in milk is about 0. 3 mg. Together with the basal iron losses of 0. 8 mg, the total iron requirements during the lactation period amount to 1. 1 mg/day.

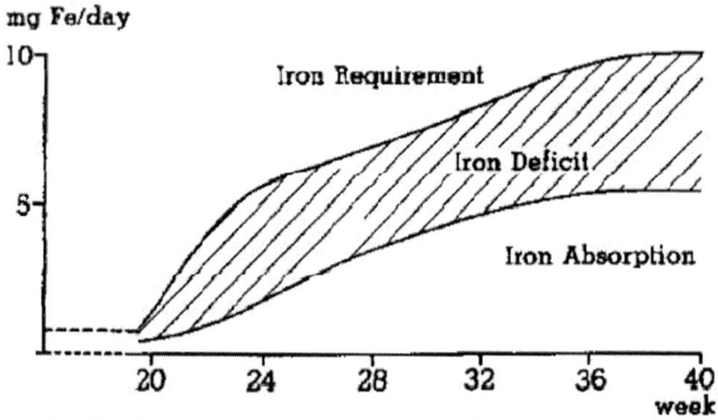

Fig. Daily iron requirements and daily dietary iron absorption in pregnancy

Early in pregnancy there are marked hormonal, haemodynamic, and haematologic changes. There is, for example, a very early increase in the plasma volume, which has been used to explain the physiologic anaemia of pregnancy observed also in iron-replete women. The primary cause of this phenomenon, however, is more probably an increased ability of the haemoglobin to deliver oxygen to the tissues (foetus). This change is induced early in pregnancy by increasing the content of 2, 3-diphospho-D-glycerate in the erythrocytes, which shifts the hemoglobin-oxygen dissociation curve to the right. The anaemia is a consequence of this important adaptation and is not primarily a desirable change, for example, to improve placental blood flow by reducing blood viscosity.

Another observation has likewise caused some confusion about the rationale of giving extra iron routinely in pregnancy. In extensive studies of pregnant women, there is a U-shaped relationship between various pregnancy complications and the haemoglobin level (i. e., there are more complications at both low and high levels). There is nothing to indicate, however, that high haemoglobin levels (within the normal non-pregnant range) per se have any negative effects. The haemoglobin increase is caused by pathologic hormonal and hemodynamic changes induced by an increased sensitivity to angiotensin II that occurs in some pregnant women, leading to a reduction in plasma volume, hypertension, and toxaemia of pregnancy.

Pregnancy in adolescents presents a special problem because iron is needed to cover the requirements of growth. In countries with very early marriage, a girl may get pregnant before menstruating. The additional iron requirements for growth of the mother are then very high and the iron situation is very serious.

In summary, the marked physiologic adjustments occurring in pregnancy are not sufficient to balance its very marked iron requirements, and the pregnant woman has to rely on her iron stores, if present. The composition of the diet has not been adjusted to the present low-energy-demanding lifestyle in industrialised countries. This is probably the main cause of the critical iron-balance situation in pregnancy today, that is due to absent or insufficient iron stores in women before they get pregnant. The unnatural necessity to give extra nutrients such as iron and folate to otherwise healthy pregnant women should be considered in this perspective.

Iron Supplementation and Fortification

The prevention of iron deficiency has become even more urgent in recent years with the accumulation of evidence strongly suggesting a relationship between even mild iron deficiency and brain development and especially with the observation that functional defects affecting learning and behaviour cannot be reversed by giving iron later on. As mentioned, iron deficiency is common both in developed and in developing countries. Great efforts have been made by WHO to develop methods to combat iron deficiency.

Iron deficiency can generally be combated by one or more of the following three strategies: 1) iron supplementation (i. e., giving iron tablets to certain target groups such as pregnant women and preschool children); 2) iron fortification of certain foods, such as flour; and 3) food and nutrition education to improve the amount of iron absorbed from the diet by increasing the intake of iron and especially by improving the bio-availability of the dietary iron.

Several factors determine the feasibility and effectiveness of different strategies, such as the health infrastructure of a society, the economy, access to suitable vehicles for iron fortification, etc. The solutions are therefore often quite different in developing and developed countries. There is an urgency to obtain knowledge about the feasibility of different methods to improve iron nutrition and to apply present knowledge. In addition, initiation of local activities should be stimulated while actions from governments are awaited.

The Evidence for Estimating the Recommended Nutrient Intake for Iron

To translate physiologic iron requirements, given in Table, into dietary iron requirements, the bio-availability of iron in different diets must be calculated. It is therefore necessary to choose an iron status where the supply of iron to the erythrocyte precursors and other tissues starts to be compromised. A state of iron-deficient erythropoiesis occurs when iron can no longer be mobilised from iron stores; iron can no longer be mobilised when stores are almost completely empty. A reduction then occurs, for example, in the concentration of haemoglobin and in the average content of haemoglobin in the erythrocytes (a reduction in mean corpuscular haemoglobin).

At the same time the concentration of transferrin in the plasma increases because of an insufficient supply of iron to liver cells. These changes were recently shown to occur rather suddenly at a level of serum ferritin of £15 μg/l. A continued negative iron balance will further reduce the level of haemoglobin. Symptoms related to iron deficiency are less related to the haemoglobin level and more to the fact that there is a compromised supply of iron to tissues.

The bio-availability of iron in meals consumed in countries with a Western-type diet has been measured by using different methods. Numerous single-meal studies have shown absorption of non-heme iron ranging from 5 percent to 40 percent. Attempts have also been made to estimate the bio-availability of dietary iron in populations consuming Western-type diets by using indirect methods (e. g., calculation of the coverage of iron requirements in groups of subjects with known dietary intake). Such estimations suggest that in borderline iron-deficient subjects the bio-availability from good diets may reach a level around 14-16 percent (15 percent relates to subjects who have a serum ferritin value of <15μg/l or a reference dose absorption of 56. 5 percent).

Recently, direct measurements were made of the average bio-availability of iron in different Western-type diets. Expressed as total amounts of iron absorbed from the whole diet, it was found that 53. 2 μg Fe/kg/day could be absorbed daily from each of the two main meals of an experimental diet which included ample amounts of meat or fish. For a body weight of 55 kg and an iron intake of 14 mg/day, this corresponds to a bio-availability of 21 percent in subjects with no iron stores and an iron-deficient erythropoiesis. A diet common among women in Sweden contained smaller portions of meat and fish, higher phytates, and some vegetarian meals each week was found to have a bio-availability of 12 percent. Reducing the intake of meat and fish further will reduce the bio-availability to about 10 percent (25μg Fe/kg/day). In vegetarians the bio-availability is usually low because of the absence of meat and fish and a high intake of phytate and polyphenols. An average good Western-type whole diet has a bio-availability of about 15 percent but for common diets, especially among women, the bio-availability is around 12 percent or even 10 percent.

In countries or for certain groups in a population with a very high meat intake, the bio-availability may rather be around 18 percent. In Western countries, a high bio-availability is mainly associated with a high meat intake, a high intake of ascorbic acid with meals, a low intake of phytate-rich cereals, and no coffee or tea within 2 hours of the main meals. Table shows examples of diets with different iron bio-availability.

Table: Examples of diet with different iron bio-availability

Type of diet	Bio-availability μg/kg/day
Preagricultural ancestors Plant/ animal subsistence: 65/35	
Very high meat and ascorbic acid	150
Very high meat in 2 main meals daily and high ascorbic acid (theoretical)	75
High meat/fish in 2 main meals daily	66. 7
Moderate meat/fish in 2 main meals daily	53. 2
Moderate meat/fish in 2 main meals daily, low phytate and calcium	42. 3
Meat/fish in 60% of 2 main meals daily, high phytate and calcium	31. 4
Lower meat intake, high phytate. Often one main meal	25
Meat/fish negligible, high phytate. High tannin, low ascorbic acid	15

Table: Translation of bio-availability expressed as amount of iron absorbed into percent absorbed for two levels of iron intake

Bio-availability, μg/kg/day	Absorption as mg Feat no iron stores in women of 55 kg body weight	Bio-availability %	
		15 mg	17 mg
150	8. 25	55	48. 8
75	4. 13	27. 5	24. 4
66. 7	3. 67	24. 5	21. 8
53. 2	2. 93	19. 5	17
42. 3	2. 32	15. 5	13. 5
31. 4	1. 73	11. 5	10
25	1. 38	9. 2	8. 2
15	0. 83	5. 5	4. 7

MAGNESIUM

Functions of Magnesium

The human body contains about 760 mg of magnesium at birth, approximately 5 g at age 4-5 months, and 25 g when adult (1-3). Of the body's magnesium, 30-40 percent is found in muscles and soft tissues, 1 percent is found in extracellular fluid, and the remainder is in the skeleton, where it accounts for up to 1 percent of bone ash.

Soft tissue magnesium functions as a co-factor of many enzymes involved in energy metabolism, protein synthesis, RNA and DNA synthesis, and maintenance of the electrical potential of nervous tissues and cell membranes. Of particular importance with respect to the pathologic effects of magnesium depletion is the role of this element in regulating potassium fluxes and its involvement in the metabolism of calcium. Magnesium depletion depresses both cellular and extracellular potassium and exacerbates the effects of low-potassium diets on cellular potassium content. Muscle potassium becomes depleted as magnesium deficiency develops, and tissue repletion of potassium is virtually impossible unless magnesium status is restored to normal. Low plasma calcium develops frequently as magnesium status declines. It is not clear whether this occurs because parathyroid hormone release is inhibited or, more probably, because of a reduced sensitivity of the bone to parathyroid hormone, thus restricting withdrawal of calcium from the skeletal matrix.

Between 50 percent and 60 percent of body magnesium is located within bone, where it is thought to form a surface constituent of the hydroxyapatite (calcium phosphate) mineral component. Initially much of this magnesium is readily exchangeable with serum and therefore represents a moderately accessible magnesium store, which can be drawn on in times of deficiency. However, the proportion of bone magnesium in this exchangeable form declines significantly with increasing age.

Significant increases in bone mineral density of the femur have been associated positively with rises in erythrocyte magnesium when the diets of subjects with gluten-sensitive enteropathy were fortified with magnesium. Little is known of other roles for magnesium in skeletal tissues.

Origins and Effects of Magnesium Deficiency

Pathologic effects of primary nutritional deficiency of magnesium occur infrequently in infants but are even less common in adults unless a relatively low magnesium intake is accompanied by prolonged diarrhoea or excessive urinary magnesium losses. Susceptibility to the effects of magnesium deficiency rises when demands for magnesium increase markedly with the resumption of tissue growth during rehabilitation from general malnutrition. Studies have shown that a decline in urinary magnesium excretion during protein-energy malnutrition (PEM) is accompanied by a reduced intestinal absorption of magnesium. The catch-up growth associated with recovery from PEM is achieved only if magnesium supply is increased substantially.

Most of the early pathologic consequences of magnesium depletion are neurologic or neuromuscular defects, some of which probably reflect the influence of the element on potassium flux within tissues. Thus, a decline in magnesium status produces anorexia, nausea, muscular weakness, lethargy, staggering, and, if deficiency is prolonged, weight loss. Progressively increasing with the severity and duration of depletion are manifestations of hyperirritability, hyperexcitability, muscular spasms, and tetany, leading ultimately to convulsions. An increased susceptibility to audiogenic shock is common in experimental animals. Cardiac arrhythmia and pulmonary oedema frequently have fatal consequences. It has been suggested that a sub-optimal magnesium status may be a factor in the aetiology of coronary heart disease and hypertension but additional evidence is needed.

Dietary Sources, Absorption, and Excretion of Magnesium

Dietary deficiency of magnesium of a severity sufficient to provoke pathologic changes is rare. Magnesium is widely distributed in plant and animal foods, and geochemical and other environmental variables rarely have a major influence on its content in foods. Most green vegetables, legume seeds, peas, beans, and nuts are rich in magnesium, as are some shellfish, spices, and soya flour, all of which usually contain more than 500 mg/kg fresh weight. Although most unrefined cereal grains are reasonable sources, many highly refined flours, tubers, fruits, and fungi and most oils and fats contribute little dietary magnesium (<100 mg/kg fresh weight). Corn flour, cassava and sago flour, and polished rice flour have an extremely low magnesium content. Table presents representative data for the dietary magnesium intakes of infants and adults.

Stable isotope studies with ^{25}Mg and ^{26}Mg indicate that between 50 percent and 90 percent of the labelled magnesium from maternal milk and infant formula can be absorbed by infants. Studies with adults consuming conventional diets show that the efficiency of magnesium absorption can vary greatly depending on magnesium intake. In one study 25 percent of magnesium was absorbed when magnesium intake was high compared with 75 percent when intake was low. During a 14-day balance study a net absorption of 52 ± 8 percent was recorded for 26 adolescent females consuming 176 mg magnesium daily.

Although this intake is far below the US recommended dietary allowance (RDA) for this age group (280 mg/day), magnesium balance was still positive and averaged 21 mg/day. This provided one of several sets of data illustrating the homeostatic capacity of the body to adapt to a wide variety of ranges in magnesium intake. Magnesium absorption appears to be greatest within the duodenum and ileum and occurs by both passive and active processes.

High intakes of dietary fibre (40-50 g/day) lower magnesium absorption. This is probably attributable to the magnesium-binding action of phytate phosphorus associated with the fibre. However, consumption of phytate- and cellulose-rich products (usually containing high concentrations of magnesium) increases magnesium intake, which often compensates for the decrease in absorption. The effects of dietary components such as phytate on magnesium absorption are probably critically important only at low magnesium intake. There is no consistent evidence that modest increases in the intake of calcium, iron, or manganese affect magnesium balance. In contrast, high intakes of zinc (142 mg/day) decrease magnesium absorption and contribute to a shift toward negative balance in adult males.

The kidney has a very significant role in magnesium homeostasis. Active reabsorption of magnesium takes place in the loop of Henle in the proximal convoluted tubule and is influenced by both the urinary concentration of sodium and probably by acid-base balance. The latter relationship may well account for the observation from Chinese studies that dietary changes which result in increased urinary pH and decreased titratable acidity also reduce urinary magnesium output by 35 percent despite marked increases in dietary magnesium input for vegetable protein diets. Several studies have now shown that dietary calcium intakes in excess of 2600 mg/day, particularly if associated with high sodium intakes, contribute to a shift toward negative magnesium balance or enhance its urinary output.

Criteria for Assessing Magnesium Requirements and Allowances

In view of the recent conclusion that many estimates of dietary requirements for magnesium were "based upon questionable and insufficient data", a closer examination is needed of the value of biochemical criteria for defining the adequacy of magnesium status. Attention could be paid to the effects of changes in magnesium intake on urinary magnesium-creatinine ratios), the relationships between serum magnesium-calcium and magnesium-potassium concentrations, and other functional indicators of magnesium status.

Estimated Allowances of Magnesium

The scarcity of studies from which to derive estimates of dietary allowances for magnesium has been emphasised by virtually all the agencies faced with this task. One United Kingdom agency commented particularly on the scarcity of studies with young subjects, and circumvented the problem of discordant data from work with adolescents and adults by restricting the range of studies considered. Using experimental data virtually identical to those used for a detailed critique of the basis for US estimates, the Scientific Committee for Food of the European Communities did not propose magnesium allowances because of inadequate data. Instead, they offered an acceptable range of intakes for adults of 150-500 mg/day and described a series of quasi-PRI values for specific age groups, including an increment of 30 percent to allow for individual variations in growth. Statements of acceptable intakes leave uncertainty as to the extent of overestimation of derived recommended intakes.

It is questionable whether more reliable estimates of magnesium requirements can be made until data from balance studies are supported by the use of biochemical indexes of adequacy that could reveal the development of manifestations of sub-optimal status. Such indexes have been examined, in their studies of the metabolic significance of magnesium depletion during PEM.

A loss of muscle and serum magnesium resulted if total body magnesium retention fell below 2 mg/kg/day and was followed by a fall in the myofibrillar nitrogen-collagen ratio of muscle and a fall in muscle potassium content. Repletion of tissue magnesium status preceded a threefold increase in muscle potassium content. It accelerated by 7-10 days at the rate of recovery of muscle mass and composition initiated by restitution of nitrogen and energy supplies to infants previously deficient.

Neurologic signs such as hyper-irritability, apathy, tremors, and occasional ataxia accompanied by low concentrations of potassium and magnesium in skeletal muscle and strongly negative magnesium balances were reported by many other studies of protein calorie deficiency in infants. Particularly noteworthy is evidence that all these effects are ameliorated or eliminated by increased oral magnesium, as were specific anomalies in the electrocardiographic T-wave profiles of such malnourished subjects. Evidence that the initial rate of growth at rehabilitation is influenced by dietary magnesium intake indicates the significance of this element for those involved in the aetiology of the PEM syndromes.

Regrettably, detailed studies have yet to be carried out to define the nature of changes resulting from a primary deficiency of dietary magnesium. Definition of magnesium requirements must continue to be based on the limited information provided by balance techniques, which give little or no indications of responses to inadequacy in magnesium supply which may induce covert pathologic changes. Reassurance must thus be sought from the application of dietary standards for magnesium in communities consuming diets differing widely in magnesium content. The inadequate definition of lower acceptable limits of magnesium intakes raises concern in communities or individuals suffering malnutrition or from a wider variety of nutritional or other diseases which influence magnesium metabolism adversely.

Derivation of Allowances for Magnesium

The infrequency with which magnesium deficiency develops in human-milk-fed infants implies that the content and physiologic availability of magnesium in human milk meets the infants' requirements. The intake of maternal milk from exclusively human-milk-fed infants 1-10 months of age ranges from 700 to 900 g/day in both industrialised and developing countries (53). If the magnesium content of milk is assumed to be 29 mg/l, the intake from milk is 20-26 mg/ day, or approximately 0. 04 mg/kcal.

The magnesium in human milk is absorbed with substantially greater efficiency (about 80-90 percent) than that of formula milks (about 55-75 percent) or solid foods (about 50 percent) (56), and such differences must be taken into account when comparing differing dietary sources. For example, a daily intake of 23 mg from maternal milk probably yields 18 mg available magnesium, a quantity similar to that of the 36 mg or more suggested as meeting the requirements of young infants given formula or other foods.

An indication of a likely requirement for magnesium at other ages can be derived from studies of magnesium-potassium relationships in muscle (58) and the clinical recovery of young children rehabilitated from malnutrition with or without magnesium fortification of therapeutic diets. 12 mg magnesium/day was not sufficient to restore positive magnesium balances, serum magnesium content, or the magnesium and potassium contents of muscle of children undergoing PEM rehabilitation.

Muscle potassium was restored to normal by 42 mg magnesium/day but higher intakes of dietary magnesium, up to 160 mg/day, were needed to restore muscle magnesium to normal. Although these studies show clearly that magnesium synergised growth responses resulting from nutritional rehabilitation, they also indicated that rectification of earlier deficits of protein and energy was a pre-requisite to initiation of this effect of magnesium.

The secondary significance of magnesium accelerating clinical recovery from PEM. They indicate that prolonged consumption of diets low in protein and energy and with a low ratio (<0. 02) of magnesium (in milligrams) to energy (in kilocalories) can induce pathologic changes which respond to increases in dietary magnesium supply.

It is noteworthy that of the balance trials intended to investigate magnesium requirements, none has yet included treatments with magnesium-energy ratios of <0. 04 or induced pathologic responses.

The relationship Mg = (kcal x 0. 0099) - 0. 0117 (SE ± 0. 0029) holds for many conventional diets (59). Some staple foods in common use have very low magnesium contents; cassava, sago, corn flour or cornstarch, and polished rice all have low magnesium-energy ratios (0. 003-0. 02). Their use in bulk merits appraisal of total dietary magnesium content.

It has been reported, with increasing frequency, that high percents (e. g., <70 percent) of individuals from some communities in Europe have magnesium intakes substantially lower than estimates of magnesium requirements derived principally. Such reports emphasise the need for reappraisal of estimates for reasons.

The estimates submitted by this Consultation must be regarded as provisional. Until additional data become available, these estimates reflect consideration of anxieties that previous recommendations for magnesium are overestimates.

They make greater allowance for developmental changes in growth rate and in protein and energy requirements. In reconsidering data cited in previous reports, particular attention has been paid to balance data suggesting that the experimental conditions established have provided reasonable opportunity for the development of equilibrium during the investigation.

Recommended magnesium intakes are presented in Table together with indications of the relationships of each recommendation to relevant estimates of the average requirements for dietary protein, and energy.

Table: Recommended nutrient intakes for magnesium (Mg) in milligrams (mg)

	Assumed body weight kg[b]	RNI		Relative intake ratios	
Age Group[a]		**mg/day**	**Mg /kg**	**Mg/ g protein**	**Mg/ kcal/day**
Infants and children					
0-6 months					
Human-milk fed	6	26	0	2. 5	0. 05
Formula fed	6	36	6. 0	2. 9	0. 06
7-12 months	9	54	6. 0	3. 9	0. 06
1-3 years	12	60	5. 5	4. 0	0. 05
4-6 years	19	76	4. 0	3. 9	0. 04
7-9 years	25	100	4. 0	3. 7	0. 05
Adolescents,					
10-18 years					
Females	49	220	4. 5	5. 2	0. 10
Males	51	230	3. 5	5. 2	0. 09
Adults,					
19-65 years					
Females	55	220	4. 0	4. 8	0. 10
Males	65	260	4. 0	4. 6	0. 10
65+ years					
Females	54	190	3. 5	4. 1	0. 10
Males	64	224	3. 5	4. 1	0. 09

1. No increment for pregnancy; 50 mg/day increment for lactation.
2. Assumed body weights of age groups derived by interpolation.
3. Intake per gram of recommended protein intake for age of subject.
4. Intake per kilocalorie estimated average requirement.

The detailed studies of magnesium economy during malnutrition and subsequent therapy, with or without magnesium supplementation, provide reasonable grounds that the dietary magnesium recommendations derived herein for young children are realistic.

Data for other ages are more scarce and are confined to magnesium balance studies. Some have paid little attention to the influence of variations in dietary magnesium content and of the effects of growth rate before and after puberty on the normality of magnesium-dependent functions.

It is assumed that during pregnancy the foetus accumulates 8 mg and foetal adnexa accumulate 5 mg magnesium. If it is assumed that this dietary magnesium is absorbed with 50 percent efficiency, the 26 mg required over a pregnancy of 40 weeks (0. 09 mg/day) can probably be accommodated by adaptation. A lactation allowance of 50-55 mg/day for dietary magnesium is made for the secretion of milk containing 25-28 mg magnesium.

An absorption efficiency of 50 percent is assumed for all solid diets; data are not sufficient to allow for the adverse influence of phytic acid on magnesium absorption from high-fibre diets or from diets with a high content of pulses. Not surprisingly, few of the representative dietary analyses presented in Table fail to meet these allowances. The few exceptions, deliberately selected for inclusion, are the marginal intakes (232 ± 62 mg) of the 168 women of Changle county and the lower intake (190 ± 59 mg) of 147 women surveyed from Tuoli county China.

Upper Tolerable Limits of Magnesium Intake

Magnesium from dietary sources is relatively innocuous. Contamination of food or water supplies with magnesium salt has been known to cause hypermagnesemia, nausea, hypotension, and diarrhoea. Intakes of 380 mg magnesium as magnesium chloride have produced such signs in women. Upper limits of 65 mg for children ages 1-3 years, 110 mg for 4-10 years, and 350 mg for adolescents and adults are suggested as tolerable limits for the content of soluble magnesium in foods and drinking water.

Relationships to Previous Estimates

The recommended intakes for infants ages 0-6 months take account of differences in the physiologic availability of magnesium from maternal milk as compared with infant formulas or solid foods. With the exception of the Canadian RNI estimates, which are 20 mg/day for ages 0-4 months and 32 mg/day for ages 5-12 months, other national estimates recommend intakes as RDAs or RNIs which substantially exceed the capacity of the lactating mother to supply magnesium for her offspring.

Recommendations for other ages are based subjectively on the absence of any evidence that magnesium deficiency of nutritional origin has occurred after consumption of a range of diets sometimes supplying considerably less than the US RDA or the UK RNI recommendations based on estimates of average magnesium requirements of 3. 4-7 mg/kg body weight.

The recommendations submitted herewith assume that demands for magnesium plus a margin of approximately 20 percent (to allow for methodologic variability) are probably met by allowing approximately 3. 5-5 mg/kg from pre-adolescence to maturity. This assumption yields estimates virtually identical to those for Canada. Expressed as magnesium allowance (in milligrams) divided by energy allowance (in kilocalories) (the latter based upon energy recommendations from UK estimates, all the recommendations of Table exceed the provisionally estimated critical minimum ratio of 0. 02.

It is appreciated that magnesium demand probably declines in late adulthood as requirements for growth diminish. However, it is reasonable to expect that the efficiency with which magnesium is absorbed declines in elderly subjects. It may well be that the recommendations are overgenerous for elderly subjects, but data are not sufficient to support a more extensive reduction than that indicated.

SELENIUM

THE ROLE OF SELENIUM IN HUMAN

Our understanding of the significance of selenium in the nutrition of human subjects has grown rapidly during the past 20 years. Demonstrations of its essentiality to rats and farm animals were followed by appreciation that the development of selenium-responsive diseases often reflected the distribution of geochemical variables which restricted the entry of the element from soils into food chains.

Table: A selection of characterised selenoproteins

Protein	Selenocysteine residues	Tissue distribution
Cytosolic glutathione peroxidase (GSHPx)	1	All, including thyroid
Phospholipid hydroperoxide GSHPx	1	All, including thyroid
Gastrointestinal GSHPx	1	Gastrointestinal tract
Extracellular GSHPx	1	Plasma, thyroid
Thioredoxin reductase	1 or 2	All including thyroid
Iodothyronine-deiodinase (type 1)	1	Liver, kidneys, and thyroid
Iodothyronine-deiodinase (type 2)	1	Central nervous system (CNS), and pituitary
Iodothyronine-deiodinase (type 3)	1	Brown adipose tissue, CNS, and placenta
Selenoprotein P	10	Plasma
Selenoprotein W	1	Muscle
Sperm capsule selenoprotein	3	Sperm tai

Such findings were the stimulus to in-depth investigations of the regional relevance of selenium in human nutrition. These studies have now yielded an increased understanding of the complex metabolic role of this trace nutrient. Selenium has been implicated in the protection of body tissues against oxidative stress, maintenance of defences against infection, and modulation of growth and development.

The selenium content of normal adult humans can vary widely. Values from 3 mg in New Sealanders to 14 mg in some Americans reflect the profound influence of the natural environment on the selenium contents of soils, crops, and human tissues. Approximately 30 percent of tissue selenium is contained in the liver, 15 percent in kidney, 30 percent in muscle, and 10 percent in blood plasma.

Much of tissue selenium is found in proteins as seleno-analogues of sulphur amino acids; other metabolically active forms include selenotrisulphides and other acid-labile selenium compounds. At least 15 selenoproteins have now been characterised. Examples are given in Table.

Functionally, there appear to be at least two distinct families of selenium-containing enzymes. The first includes glutathione peroxidases and thioredoxin reductase, which are involved in controlling tissue concentrations of highly reactive oxygen-containing metabolites.

These metabolites are essential at low concentrations for maintaining cell-mediated immunity against infections but highly toxic if produced in excess. The role of selenium in the cytosolic enzyme glutathione peroxidase (GSHPx) was first illustrated in 1973. During stress, infection, or tissue injury, selenoenzymes may protect against the damaging effects of hydrogen peroxide or oxygen-rich free radicals. This family of enzymes catalyses the destruction of hydrogen peroxide or lipid hydroperoxides according to the following general reactions:

H_2O_2 + 2GSH àà 2H_2O + GSSG

ROOH + 2GSH àà ROH + H_2O + GSSG

Where GSH is glutathione and GSSG is its oxidised form. At least four forms of GSHPx exist; they differ both in their tissue distribution and in their sensitivity to selenium depletion. The GSHPx enzymes of liver and blood plasma fall in activity rapidly at early stages of selenium deficiency. In contrast, a form of GSHPx associated specifically with phospholipid-rich tissue membranes is preserved against selenium deficiency and is believed to have broader metabolic roles (e. g., in prostaglandin synthesis). In concert with vitamin E, selenium is also involved in the protection of cell membranes against oxidative damage.

The selenoenzyme thioredoxin reductase is involved in disposal of the products of oxidative metabolism. It contains two selenocysteine groups per molecule and is a major component of a redox system with a multiplicity of functions, among which is the capacity to degrade locally excessive and potentially toxic concentrations of peroxide and hydroperoxides likely to induce cell death and tissue atrophy.

Another group of selenoproteins is essential in the conversion of thyroxin, or tetraiodothyronine (T_4), to its physiologically active form, triiodothyronine (T_3). Three types of these iodothyronine deiodinases, differing both in tissue distribution and sensitivity to selenium deficiency, have been characterised. The consequences of a low selenium status on physiologic responses to a shortage of iodine are complex.

The influence of a loss of selenium-dependent iodothyronine deiodinase differs in its severity depending on whether a target tissue needs a preformed supply of T_3 (e. g., via plasma) or whether, as with the brain, pituitary gland, and placenta, it can rely upon local synthesis of T_3 from T_4. Despite this, marked changes in the T_3-T_4 ratio as a consequence of a reduced selenium status (when iodine supplies are also marginal) indicate the modifying influence of selenium on thyroid hormone balance in both animal models and human subjects. Their possible significance can be anticipated from the fact that whereas thyroid weights increase typically by 50 percent in rats offered an iodine-deficient diet, thyroid weight is increased 154 percent by diets concurrently deficient in both selenium and iodine.

Between 60 percent and 80 percent of selenium in human plasma is accounted for by a well-characterised fraction designated selenoprotein P, the function of which has yet to be determined. It is thought to be a selenium storage protein because there is limited evidence that it also has an antioxidant role. At least 10 other selenoproteins exist, including one which is a component of the mitochondrial capsule of sperm cells, damage to which may account for the development of sperm abnormalities during selenium deficiency. Other aspects of the function and metabolism of selenium are reviewed elsewhere.

Selenium Deficiency

Non-endemic Deficiencies of Selenium

Biochemical evidence of selenium depletion (e. g., a decline in blood GSHPx activity) is not uncommon in subjects maintained on parenteral or enteral feeding for long periods. Blood selenium values declining to one-tenth of normal values have been reported when the selenium content of such preparations has not been maintained by fortification.

Low selenium contents of some commercial formulas for infants resulting in a fall in daily selenium intake to approximately 0. 5 μg/day have been shown to strongly exacerbate the fall in serum selenium and GSHPx activity normally experienced from 2 to 8 months of age even in human-milk-fed infants typically receiving threefold higher selenium intakes. The importance of maintaining trace element levels in such preparations was reviewed elsewhere.

Clinical manifestations of deficiency arising from such situations are uncommon and poorly defined. They include muscular weakness and myalgia with, in several instances, the development of congestive heart failure. In at least one instance such pathologic signs have developed as a consequence of a generally inadequate diet providing selenium at less than 10 μg/day. The 2-year-old subject recovered rapidly after selenium administration. With this last exception, virtually all of the above reports describe observations with subjects under close medical supervision. This may well be relevant to the scarcity of consistent pathologic findings.

Keshan Disease

Keshan disease was first described in Chinese medical literature more than 100 years ago, but not until 40 years after its widespread occurrence in 1935 was it discovered that selenium deficiency was an important factor in its aetiology. Endemic in children aged 2-10 years and in women of childbearing age, this disease has a geographic distribution covering localities from northeast to southwest China.

Typical manifestations are fatigue after even mild exercise, cardiac arrhythmia and palpitations, loss of appetite, cardiac insufficiency, cardiomegaly, and congestive heart failure. Pathologic changes include a multifocal myocardial necrosis and fibrosis. The coronary arteries are essentially unaffected. Ultrastructural studies show that membranous organelles, such as mitochondria or sarcolemma, are affected earliest. The disease has a marked seasonal fluctuation in incidence and may appear after only 3 months exposure to conditions in localities known to be associated with a high risk of myocarditis. Once the disease is established, selenium is of little or no therapeutic value. Prophylaxis consisting of oral administration of selenium 3 months before the periods of highest anticipated risk is highly effective.

Although geographic similarities in the distribution of Keshan disease and the selenium and vitamin E-responsive white muscle disease in animals first prompted successful investigation of the relevance of a low selenium status, evidence has grown steadily that the disease is multifactorial in origin. The strongest suspicions have fallen on the development of a viral myocarditis probably attributable to enhancement of the virulence of a coxsackievirus during its passage through selenium-deficient host tissues.

Although other nutritional variables such as a marginal vitamin E status may also be involved, the finding of extremely low selenium contents in staple crops of affected areas and convincing demonstrations of the prophylactic effectiveness of selenium administration leave no doubt that selenium deficiency is the primary factor.

Recent studies indicate that geochemical variables have an important influence on the distribution of Keshan disease. Acid soils high in organic matter and iron oxide content appear to be responsible for fixing selenium in forms which are poorly absorbed by staple crops which, in the instance of cereal grains, typically have a selenium content of less than 0. 01 μg/g. A similar geochemical background is believed to be associated with reports of selenium-responsive disorders resembling Keshan disease in the Transbaikalia region of south Siberia. Dietary intakes of selenium are inadequate to maintain blood GSHPx activity; biochemical indicators of tissue peroxidative damage are elevated until selenium therapy is initiated.

Kaschin-beck Disease

A selenium-responsive bone and joint disease (osteoarthropathy) has been detected in children aged 5-13 years in China and less extensively in south-east Siberia. The disease is characterised by joint necrosis - epiphyseal degeneration of the arm and leg joints resulting in structural shortening of the fingers and long bones with consequent growth retardation and stunting.

Although not identical to Keshan disease, Kaschin-Beck disease also occurs in areas where the availability of soil selenium for crop growth is low. The selenium contents of hair and of whole blood are abnormally low and the blood content of GSHPx is reduced. Although it is ameliorated by selenium therapy, other factors such as the frequent presence of mycotoxins in cereal grains grown in the area may be involved. A spontaneous decrease in incidence from 1970 (44 percent) to 1980 (14 percent) to 1986 (1 percent) has been attributed to general improvements in the nutritional status of Chinese rural communities.

Selenium Status and Susceptibility to Infection

As stated earlier, the expressions of the cardiac lesions of Keshan disease probably involve not only the development of selenium deficiency but also the presence of a Coxsackie virus (BA) infection. Animal studies have confirmed that selenium-deficient mice infected with Coxsackie virus (CVB/0) were particularly susceptible to the virus. These studies also illustrated that passage of the virus through selenium-deficient subjects enhanced its virulence. Myocarditic virulence developed even in strains such as CVB/0 which normally were not myopathogenic.

The enhancement of virulence in this RNA virus involves modifications to the nucleotide sequence of the phenotype. These modifications were maintained and expressed even during subsequent passage through animals with normal selenium status.

Enhancing the virulence of a virus with a selenium deficiency (resulting either from a nutritional challenge or an increased metabolic demand on tissue selenium depots) appears not to be unique to the Coxsackie viruses. The early pre-clinical stages of development of human immunodeficiency virus (HIV) infection are accompanied by a very marked decline in plasma selenium. Sub-clinical malnutrition assumes increased significance during the development of acquired immune deficiency syndrome (AIDS).

However, for the nutrients affected, there are strong indications that only the extent of the decline in selenium status has predictive value with respect to both the rate of development of AIDS and its resulting mortality. The virulence of other RNA viruses such as hepatitis B and those associated with the development of haemolytic anaemias are enhanced similarly by a decline in selenium status. The mechanisms underlying these effects are not yet resolved. There are indications that the loss of protective antioxidant functions dependent on selenium and vitamin E are both involved and that the resulting structural changes in viral nucleotide sequences are reproducible and appear to provoke additional selenoprotein synthesis. It is suspected that this further depletes previously diminished pools of physiologically available selenium and accelerates pathologic responses.

Whatever mechanisms are involved, further understanding is needed of the influence of selenium status on susceptibility to viral diseases ranging from cardiomyopathies to haemolytic anaemias. The relationship already illustrates the difficulty of defining nutritional essentiality for nutrients which may primarily maintain defences against infection. Studies of the effects of selenium deficiency in several experimental animal species have shown that the microbicidal activity of blood neutrophils is severely impaired even though phagocytic activity remains unimpaired. The complexity of species differences in the influence of selenium status on the effectiveness of cell-mediated immune processes is summarised elsewhere.

The possibility that increased intakes of selenium might protect against the development of cancer in humans has generated great interest. However, a number of epidemiologic studies have now been reported which show no relationship between selenium and cancer risk. Moreover, an analysis of the relationship between selenium and cancer suggests that "the question of whether selenium protects against cancer is still wide open".

An increased intake of selenium appears to stimulate tumorigenesis in some animal models of pancreatic and skin cancer. In contrast, the protective effect of higher exposures to selenium observed in several animal studies, together with small but statistically significant differences in selenium blood plasma levels detected in some retrospective-prospective studies of subgroups of people developing cancer, explains the continuing interest in the anticarcinogenic potential of selenium. However, the results of prospective-retrospective studies had no predictive value for individuals and could have reflected non-specific influences on groups. The association between low selenium intake and high cancer risk, although clearly of some interest, is in need of further investigation before a conclusion can be reached.

Although a biochemical mechanism can be postulated whereby selenium could protect against heart disease by influencing platelet aggregation (through an effect on the prostacyclin-thromboxane ratio), the epidemiologic evidence linking selenium status and risk of cardiovascular disease is still equivocal.

Selenium and Thyroid Hormones

The importance of selenium for thyroid hormone metabolism is evident from changes in the T_3-T_4 ratio which develop after relatively mild selenium depletion in infants and elderly (65+ years) subjects. Decreases in the T_3-T_4 ratio indicative of decreased thyroid hormone balance have been detected when serum selenium falls below 0. 9 μmol/l. In a recent Scottish study these decreases were correlated with a decline in dietary and plasma selenium after the replacement of selenium-rich wheat from Canada and the United States with selenium-deficient wheat from European sources.

Communities noted for a high incidence of myxedematous cretinism have been found to have low plasma selenium status and GSHPx activity in addition to having low iodine status and being exposed to high thiocyanate intakes from cassava. Restoration of iodine supply, particularly if excessive, tends to induce a high peroxidative stress through the action of iodide peroxidase, the first step in iodine utilisation by the thyroid. It is postulated that necrosis and thyroid fibrosis leading to irreversible hypothyroidism result if a concurrent deficiency of selenium limits peroxide destruction by the protective action of the selenium-dependent enzymes, GSHPx and, more probably, thioredoxin reductase.

In areas where myxedematous cretinism is endemic and characterised by persistent hypothyroidism, dwarfism, and stunting, it has been recommended that attempts to introduce iodine therapy for mildly affected individuals should be preceded by an assessment of selenium status and rectification of any observed deficit. Although this suggestion is compatible with pathologic observations on hypothyroid rats differing in selenium status, its validity has yet to be assessed adequately in humans.

The Influence of Diet on Selenium Status

Environmental conditions and agricultural practices have a profound influence on the selenium content of many foods.

The wide range of selenium content of the principal food groups and the variability in the selenium content of individual foods.

This variability is exceeded only by that found in the iodine content of foods. Geographic differences in the content and availability of selenium from soils to food crops and animal products have a marked effect on the selenium status of entire communities. For example, the distribution of Keshan disease and Kaschin-Beck disease in China reflects the distribution of soils from which selenium is poorly available to rice, maise, wheat, and pasture grasses.

Cereal crop selenium contents of 3-7 ng/g are not uncommon and it has been suggested that <10 ng/g for grain selenium and <3 ng/g for water-soluble soil selenium could be used as indexes to define deficient areas. Fluctuations in the selenium status of many communities in northern Europe reflect the intrinsically low selenium content of its glacial soils and the extent to which selenium supplementation of fertilisers has been successful in increasing the selenium content of cereal grains, milk, and other animal products. Deliberate importation of cereals from areas with relatively high available selenium in soil has also occurred or been recommended in some areas of Finland, New Sealand, and the United Kingdom after steady declines in the selenium status of some communities were noted. Conversely, low-selenium grains are being selected in China, India, and Venezuela to reduce the risks of selenosis.

Comprehensive data summarising the selenium contents of staple foods are available elsewhere. Reports from the Food and Agriculture Organization of the United Nations (FAO) and the International Atomic Energy Agency (IAEA) provide representative data on daily total selenium intakes for more than 40 countries. The great influence of dietary and geographic variables on selenium status is evident from recent summaries of data describing national and regional differences for the selenium content of human and formula milks, diets, and human serum.

Absorption and Bio-availability

Selenium compounds are generally very efficiently absorbed by humans, and selenium absorption does not appear to be under homeostatic control. For example, absorption of the selenite form of selenium is greater than 80 percent whereas that of selenium as selenomethionine or as selenate may be greater than 90 percent.

Therefore, the rate-limiting step determining the overall availability of dietary selenium is not likely to be its absorption but rather its conversion within tissues to its metabolically active forms (e. g., its incorporation into GSHPx or 5'-deiodinase). A number of depletion-repletion experiments have been carried out on animals to estimate the bio-availability of selenium in human foods. Based on the restoration of GSHPx activity in depleted rats, the bio-availability of selenium in wheat is quite good, usually 80 percent or better.

The selenium in Brazil nuts and beef kidney also appears readily available (90 percent or more by most criteria). The selenium in tuna seems of lesser availability (perhaps only 20-60 percent of that from selenite) whereas the availability of selenium from certain other seafoods (shrimp, crab, and Baltic herring) is high. The selenium in a variety of mushrooms appears to be of uniformly low availability to rats.

Data on the nutritional bio-availability of selenium to humans are sparse. A supplementation study carried out on Finnish men of relatively low selenium status showed that selenate selenium was as effective as the selenium in seleniferous wheat in increasing platelet GSHPx activity. The wheat selenium, however, increased plasma selenium levels more than did selenate selenium and once the supplements were withdrawn, platelet GSHPx activity declined less in the group given wheat. This study showed the importance of estimating not only short-term availability but also long-term retention and the convertibility of tissue selenium stores into biologically active forms.

Criteria for Assessing Selenium Requirements

Levander convincingly illustrated the impracticability of assessing selenium requirements from input-output balance data because the history of selenium nutrition influences the proportion of dietary selenium absorbed, retained, or excreted. The changing equilibria when selenium intake is varied experimentally yield data which are of limited value for estimating minimal requirements. Examples are cited of estimates of selenium requirement for adults of 7. 4 and 80 μg/day derived from Chinese and United States studies, respectively. Such discrepancies reflect differences in the usual daily selenium intakes of the experimental subjects and the extent to which this was changed experimentally. This situation, not unique to selenium, emphasises the importance of basing requirement estimates on functional criteria derived from evidence describing the minimum levels of intake which, directly or indirectly, reflect the normality of selenium-dependent processes.

New opportunities for the development of biochemical indexes of selenium adequacy such as those listed in Table have yet to be exploited. Until this is done, the most suitable alternative is to monitor changes in the relationship of serum selenium to dietary selenium supply, taking advantage of its relatively constant proportionality to the fraction of serum selenium in functionally significant GSHPx.

A detailed review of 36 reports describing serum selenium values in healthy subjects indicated that they ranged from a low of 0. 52 μmol/l in Serbia to a high of 2. 5 μmol/l. It was suggested that mean values within this range derived from 7502 apparently healthy individuals should be regarded tentatively as a standard for normal reference. This survey clearly illustrated the influence of crop management on serum selenium level; in Finland and New Sealand, selenium fortification of fertilisers for cereals increased serum selenium from 0. 6 to 1. 5 μmol/l. A summary of these data in Table also includes representative mean serum selenium values within the range of 0. 15-0. 55 μmol/l reported for specific diseases known to be associated with disturbances in selenium nutrition or metabolism. These include reports from studies of Keshan disease, Kaschin-Beck disease, and specific studies of cretinism, hypothyroidism, and HIV and AIDS where clinical outcome or prognosis has been related to selenium status.

This report and the report by the World Health Organization (WHO), FAO, and IAEA use virtually identical approaches to derive their estimates of basal requirements for selenium (Se_R^{basal}). Some modification is necessary however to estimate population minimum intakes with adequate allowance for the variability (CV) associated with estimates of the average selenium intakes from the typical diets of many communities. A CV of 16 percent was assumed for the selenium conventional diets and 12. 5 percent for the milk-based diets of infants to limit the risks of inadequacy arising from unexpectedly low selenium contents.

More recent studies suggest that the variability of selenium intake from diets for which the selenium content has been predicted rather than measured may be substantially greater than estimated previously.

Recommended Selenium Intakes

Because balance techniques were shown to be inappropriate for determining selenium requirements, presented requirement estimates based on of epidemiologic evidence derived from areas of China endemic or non-endemic for Keshan disease.

These comprehensive biochemical and clinical studies showed that Keshan disease did not occur in regions where the mean intake of selenium by adult males or females was greater than 19. 1 or greater than 13. 3 μg/day, respectively. Although these intakes were sufficient to eliminate clinical evidence of myocarditis and other signs of Keshan disease, other studies showed that they were inadequate to restore erythrocyte or plasma selenium concentrations or GSHPx activities to levels indicative of reserves.

Studies with adult male subjects initially of low selenium status given a carefully monitored diet providing selenium at 11 μg/day together with supplements of selenomethionine given orally which provided 0, 10, 30, 60, or 90 μg/day. Starting at frankly deficient levels, total daily selenium intakes of above 41 μg/day were found sufficient increase plasma GSHPx substantially and to saturate plasma activity in 60-kg male subjects within 5-8 months. It was estimated that satisfactory levels of plasma selenium (>80 μmol/l) and of GSHPx (>0. 3 mmol NADPH oxidised/min/l; approximately two-thirds of plasma saturation activity) indicative of adequate selenium reserves would be attained after intakes of approximately 27 μg/day by 65-kg male subjects. Such criteria satisfying the definition of average normative requirements for selenium ($Se_R^{normative}$) have been used as the basis for calculating recommended nutrient intake (RNI) values in this report after interpolating estimates of average requirements by allowing for differences in weight and basal metabolic rate of age groups to up to 65 years and adding a 25 percent increase (2 x assumed SD) to allow for individual variability in the estimates of RNI

a. Weight (kg) interpolated from FAO/WHO.
b. Derived from WHO-FAO-IAEA values.
c. Recommended nutrient intake (RNI) derived from average + 2 x assumed standard error

The estimates of RNI for infants are compatible with estimates of the international reference range of the selenium content of breast milk, with data from an extensive international survey of breast milk selenium (WHO-IAEA [50]) and with WHO data on the milk consumption of exclusively human-milk-fed infants in developed and developing countries. Data from the WHO-IAEA survey from six countries suggest that the human milk from all countries met the RNI for infants aged 0-6 months. In two of six countries, Hungary and Sweden, the human milk selenium was marginal with respect to the RNI for infants aged 7-12 months.

SELENIUM REQUIREMENTS DURING PREGNANCY AND LACTATION

The increase of dietary selenium needed for pregnancy by factorial estimation of the likely quantity of selenium incorporated into the tissues of the foetus. Such estimates have assumed that the total products of conception amount to 4. 6-6 kg lean tissue with a protein content of approximately 18. 5-20 percent. If, as appears to be a reasonable assumption, the selenium content of this protein resembles that of a skeletal muscle, growth of these tissues could account for between 1. 0 and 4. 5 μg/day of selenium depending on whether the analyses reflect consumption of diets from a low-selenium (but non-pathogenic) environment or from a region with relatively high selenium intakes.

Typically such estimates have assumed an 80 percent absorption and utilisation of dietary selenium from which it would appear reasonable to estimate that allowing for a variability of estimates (CV 12. 5 percent) an increase of 2 μg/day would be appropriate for the second trimester and 4 μg/day would be appropriate for the third trimester of pregnancy.

As is evident from Table the selenium content of human milk is sensitive to changes in maternal dietary selenium. The increase of maternal dietary selenium needed to meet requirements for lactation has been estimated from the estimated RNI for infants aged 0-6 months and 7-12 months. It is assumed that the selenium of maternal milk is used with an efficiency of 80 percent, and a SD of 12. 5 percent is assumed. For the period 0-6 months it is estimated that the infant must receive 6 μg/day from human milk. The increase of maternal dietary selenium required to produce this will be 6 x + (2 x SD) = 9 μg/day. The corresponding increase needed to meet the infant RNI of 10 μg/day for ages 7-12 months will be 16 μg/day. Added to the non-pregnancy maternal RNI of 26 μg/

day, the total RNI for lactation during the first 6 months post-partum will be 35 μg/day and for months 7-12 will be 42 μg/day.

As implied by the data in Tables 48, 49 and 50, agricultural growing practices, geologic factors, and social deprivation enforcing the use of an abnormally wide range of dietary constituents may significantly modify the variability of dietary selenium intakes. If accumulated experience suggests that the CV of selenium intake may be 40 percent or more and tabulated rather than analysed data are used to predict the dietary content of selenium, the selenium allowance may have to be increased accordingly or assessed by using the WHO-FAO-IAEA technique.

Upper Tolerable Nutrient Level for Selenium

A comprehensive account of the clinically significant biochemical manifestations of chronic and acute intoxication from selenium arising from high concentrations in food, drinking water. The signs and symptoms of human overexposure to selenium are not well defined. Common clinical features are hair loss and structural changes in the keratin of hair and of nails, the development of icteroid skin, and gastrointestinal disturbances. An increased incidence of nail dystrophy has been associated with consumption of high-selenium foods supplying more than 900 μg/day. These foods were grown in selenium-rich (seleniferous) soil from specific areas. A positive association between dental caries and urinary selenium output under similar circumstances.

Sensitive biochemical markers of impending selenium intoxication have yet to be developed. In their absence it is suggested that the upper tolerable nutrient intake level (UL) for selenium should be set, provisionally, at 400 μg/day for adults. It is noteworthy that a maximum tolerable dietary concentration of 2 mg/kg dry diet was suggested for all classes of domesticated livestock and has proved satisfactory in use. This suggests that the proposed UL of 400 μg/day for human subjects provides a fully adequate margin of safety. The UL for children and for pregnant or lactating women has yet to be determined.

Food commodity inputs are changing rapidly and in some instances, unpredictably. Under most circumstances it will be unreasonable to expect that the often marked influence of geographic variability on the supply of selenium from cereals and meats can be taken into account. Changes in trade patterns with respect to the sources of cereals and meats are already having significant influences on the selenium nutrition of consumer communities. Such evidence fully justifies the warning to allow for a high intrinsic variability of dietary selenium content when estimating selenium requirements of populations for which the principal sources of this microelement are unknown.

ZINC

ROLE OF ZINC IN HUMAN METABOLIC PROCESSES

Zinc is present in all body tissues and fluids. The total body zinc content has been estimated to be 30 mmol (2 g). Skeletal muscle accounts for approximately 60 percent of the total body content and bone mass, with a zinc concentration of 1. 5-3 mmol/g (100-200 mg/g), for approximately 30 percent. Zinc concentration of lean body mass is approximately 0. 46 mmol/g (30 mg/g). Plasma zinc has a rapid turnover rate and it represents only about 0. l percent of total body zinc content. This level appears to be under close homeostatic control. High concentrations of zinc are found in the choroid of the eye 4. 2 mmol/g (274 μg/g) and in prostatic fluids 4. 6-7. 7 mmol/l (300-500 mg/l).

Zinc is an essential component of a large number (>300) of enzymes participating in the synthesis and degradation of carbohydrates, lipids, proteins, and nucleic acids as well as in the metabolism of other micronutrients. Zinc stabilises the molecular structure of cellular components and membranes and contributes in this way to the maintenance of cell and organ integrity. Furthermore, zinc has an essential role in polynucleotide transcription and thus in the process of genetic expression. Its involvement in such fundamental activities probably accounts for the essentiality of zinc for all life forms.

Zinc plays a central role in the immune system, affecting a number of aspects of cellular and Humoural immunity. The clinical features of severe zinc deficiency in humans are growth retardation, delayed sexual and bone maturation, skin lesions, diarrhoea, alopecia, impaired appetite, increased susceptibility to infections mediated via defects in the immune system, and the appearance of behavioural changes. The effects of marginal or mild zinc deficiency are less clear. A reduced growth rate and impairments of immune defence are so far the only clearly demonstrated signs of mild zinc deficiency in humans. Other effects, such as impaired taste and wound healing, which have been claimed to result from a low zinc intake, are less consistently observed.

Zinc Metabolism and Homeostasis

Zinc absorption is concentration dependent and occurs throughout the small intestine. Under normal physiologic conditions, transport processes of uptake are not saturated. Zinc administered in aqueous solutions to fasting subjects is absorbed efficiently (60-70 percent), whereas absorption from solid diets is less efficient and varies depending on zinc content and diet composition.

Zinc is lost from the body through the kidneys, skin, and intestine. The endogenous intestinal losses can vary from 7 mmol/day (0. 5 mg/day) to more than 45 mmol/day (3 mg/ day), depending on zinc intake. Urinary and skin losses are of the order of 7-10 mmol/day (0. 5-0. 7 mg/day) each and depend less on normal variations in zinc intake. Starvation and muscle catabolism increase zinc losses in urine. Strenuous exercise and elevated ambient temperatures could lead to losses by perspiration.

The body has no zinc stores in the conventional sense. In conditions of bone resorption and tissue catabolism, zinc is released and may be re-utilised to some extent. Human experimental studies with low-zinc diets 2. 6-3. 6 mg/day (40-55 mmol/day) have shown that circulating zinc levels and activities of zinc-containing enzymes can be maintained within normal range over several months, which highlights the efficiency of the zinc homeostasis mechanism. Controlled depletion-repletion studies in humans have shown that changes in the endogenous excretion of zinc through the kidneys, intestine, and skin and changes in absorptive efficiency are how body zinc content is maintained. The underlying mechanisms are poorly understood.

Sensitive indexes for assessing zinc status are unknown at present. Static indexes, such as zinc concentration in plasma, blood cells, and hair, and urinary zinc excretion are decreased in severe zinc deficiency. A number of conditions that are unrelated to zinc status can affect all these indexes, especially zinc plasma levels. Infection, stress situations such as fever, food intake, and pregnancy lower plasma zinc concentrations whereas, for example, long-term fasting increases it. However, on a population basis, reduced plasma zinc concentrations seem to be a marker for zinc-responsive growth reductions. Experimental zinc depletion studies suggest that changes in immune response occur before reductions in plasma zinc concentrations are apparent. So far, it has not been possible to identify zinc-dependent enzymes which could serve as early markers for zinc status.

A number of functional indexes of zinc status have been suggested, for example, wound healing, taste acuity, and dark adaptation. Changes in these functions are, however, not specific to zinc and these indexes have so far not been proven useful for identifying marginal zinc deficiency in humans.

The introduction of stable isotope techniques in zinc research has created possibilities for evaluating the relationship between diet and zinc status and is likely to lead to a better understanding of the mechanisms underlying the homeostatic regulations of zinc. Estimations of turnover rates of administered isotopes in plasma or urine have revealed the existence of a relatively small rapidly exchangeable body pool of zinc of about 1. 5-3 mmol (100-200 mg). The sise of the pool seems to be correlated to habitual dietary intake and it is reduced in controlled depletion studies.

The exchangeable zinc pool was also found to be correlated to endogenous faecal excretion of zinc and to total daily absorption of zinc. These data suggest that the sise of the exchangeable pool depends on recently absorbed zinc and that a larger exchangeable pool results in larger endogenous excretion. Changes in endogenous intestinal excretion of zinc seem to be more important than changes in absorptive efficiency for maintenance of zinc homeostasis.

DIETARY SOURCES AND AVAILABILITY OF ZINC

Lean red meat, whole-grain cereals, pulses, and legumes provide the highest concentrations of zinc 25-50 mg/kg (380-760 mmol/kg) raw weight. Processed cereals with low extraction rates, polished rice, and lean meat or meat with high fat content have a moderate zinc content 10-25 mg/kg (150-380 mmol/kg). Fish, roots and tubers, green leafy vegetables, and fruits are only modest sources of zinc <10 mg/kg (<150 mmol/kg). Separated fats and oils, sugar, and alcohol have a very low zinc content.

The utilisation of zinc depends on the overall composition of the diet. Experimental studies have identified a number of dietary factors as potential promoters or antagonists of zinc absorption. Soluble low-molecular-weight organic substances, such as amino and hydroxy acids, facilitate zinc absorption. In contrast, organic compounds forming stable and poorly soluble complexes with zinc can impair absorption. In addition, competitive interactions between zinc and other ions with similar physicochemical properties can affect the uptake and intestinal absorption of zinc. The risk for competitive interactions seems mainly to be related to high doses in the form of supplements or in aqueous solutions. However, at levels present in food and at realistic fortification levels, zinc absorption appears not to be affected, for example, by iron and copper.

Isotope studies with human subjects have identified two factors which together with the total zinc content of the diet are major determinants of absorption and utilisation of dietary zinc. The first is the content of inositol hexaphosphate (phytate) and the second is the level and source of dietary protein. Phytates are present in whole-grain cereals and legumes and in smaller amounts in other vegetables.

They have a strong potential for binding divalent cations and their depressive effect on zinc absorption has been demonstrated in humans. The molar ratio between phytates and zinc in meals or diets is a useful indicator of the effect of phytates in depressing zinc absorption. At molar ratios above the range of 6-10, zinc absorption starts to decline; at ratios above 15 absorption is typically less than 15 percent. The effect of phytate is, however, modified by the source and amount of dietary proteins consumed.

Animal proteins improve zinc absorption from a phytate-containing diet. Zinc absorption from some legume-based diets is comparable with that from animal-protein-based diets despite a higher phytate content in the former. High dietary calcium potentiated the antagonistic effects of phytates on zinc absorption in experimental studies. The results from human studies are less consistent and any effects seem to depend on the source of calcium and the composition of the diet.

The effect of zinc content and diet composition on fractional zinc absorption. The results from the total diet studies, where all main meals of a day's intake have been extrinsically labelled, show a remarkable consistency in fractional absorption despite relatively large variations in meal composition and zinc content. Thus, approximately twice as much zinc was absorbed from a non-vegetarian or high-meat diet than from a diet in rural China based on rice and wheat flour. Data are lacking on zinc absorption from typical diets of developing countries, which usually have a high phytate content.

The availability of zinc from the diet can be improved by reductions in the phytate content and inclusion of animal protein sources. Lower extraction rates of cereal grains will result in lower phytate content but at the same time the zinc content is reduced, so that the net effect on zinc supply is limited. The phytate content can be reduced by activating the phytase present in most phytate-containing foods or through the addition of microbial or fungal phytases. Phytases hydrolyse the phytate to lower inositol phosphates, resulting in an improved zinc absorption. The activity of phytases in tropical cereals such as maise and sorghum is lower than that in wheat and rye. Germination of cereals and legumes increases phytase activity and addition of some germinated flour to ungerminated maise or sorghum followed by soaking at ambient temperature for 12-24 hours can reduce the phytate content substantially. Additional reduction can be achieved by the fermentation of porridge for weaning foods or doughs for bread making. Commercially available phytase preparations could also be used but may not be economically accessible in many populations.

Populations at Risk for Zinc Deficiency

The central role of zinc in cell division, protein synthesis, and growth makes infants, children, adolescents, and pregnant women especially at risk for an inadequate zinc intake. Zinc-responsive stunting has been identified in several studies, and a more rapid body weight gain in malnourished children supplemented with zinc was reported. Other studies have failed to show a growth-promoting effect of zinc supplementation. A recent meta-analysis of 25 intervention trials comprising 1834 children under 13 years of age, with a mean duration of approximately 7 months and a mean dose of zinc of 14 mg/day (214 mmol/day), showed a small but significant positive effect of zinc supplementation on height and weight increases. The initial presence of stunting was significantly associated with an effect of zinc supplementation on height, whereas initial low plasma zinc concentrations were associated with a more pronounced effect on weight gain.

Results from zinc supplementation studies suggest that a low zinc status in children not only affects growth but is also associated with an increased risk of severe infectious diseases. Episodes of acute diarrhoea with shorter duration and less severity and reductions in incidence of diarrhoea in zinc-supplemented groups have been reported.

Other studies indicate that the incidence of acute lower respiratory tract infections and malaria may also be reduced by zinc supplementation. Prevention of sub-optimal zinc status and zinc deficiency in children by an increased intake and availability of zinc could consequently have a significant effect on child health in developing countries.

Zinc Requirements

The lack of specific and sensitive indexes for zinc status limits the possibilities for evaluating zinc requirements from epidemiologic observations. Experimental zinc repletion studies with low zinc intakes have clearly shown that the body has a pronounced ability to adapt to different levels of zinc intakes by changing the endogenous zinc losses through the kidneys, intestine, and skin (5-9,33).

The normative requirement for absorbed zinc was defined as the obligatory loss during the early phase of zinc depletion before adaptive reductions in excretion take place and was set at 1. 4 mg/day for men and 1. 0 mg/day for women. To estimate the normative maintenance requirements for other age groups, the respective basal metabolic rates were used for extrapolation. In growing individuals the rate of accretion and zinc content of newly formed tissues were used to obtain the data required for tissue growth. Similarly, the retention of zinc during pregnancy and the zinc concentration in milk at different stages of lactation were used to estimate the physiologic requirements in pregnancy and lactation.

The translation of these estimates of absorbed zinc to requirements for dietary zinc involves several considerations. First, the nature of the diet (i. e., its content of promoters and inhibitors of zinc absorption) determines the fraction of the dietary zinc that is potentially absorbable. Second, the efficiency of absorption of potentially available zinc is inversely related to the content of zinc in the diet. The review of available data from experimental zinc absorption studies of single meals or total diets resulted in a division of diets into three categories - high, moderate, and low zinc bio-availability. It was then discovered that the relationship between efficiency of absorption and zinc content differed for these diets.

Algorithms were developed and applied to the estimates of requirements for absorbed zinc to achieve a set of figures for the average individual dietary zinc requirements. The fractional absorption figures applied for the three diet categories were 50 percent, 30 percent, and 15 percent, respectively. From these estimates and from the evaluation of data from dietary intake studies, mean population intakes were identified which were deemed sufficient to ensure a low prevalence of individuals at risk of inadequate zinc intake.

Infants, Children, and Adolescents

Endogenous losses of zinc in human-milk-fed infants were assumed to be 20 μg/kg/day (0. 31mmol/kg/day) whereas 40 μg/kg/day (0. 6 mmol/kg/day) was assumed for infants fed formula or weaning foods. For other age groups an average loss of 0. 002 mmol/basal kJ (0. 57 μg/basal kcal) was derived from the estimates in adults. Estimated zinc increases for infant growth were set at 120 and 140 μg/kg/day (1. 83-2. 14 mmol/kg/day) for female and male infants, respectively, for the first 3 months. These values decrease to 33 μg/kg/day (0. 50 mmol/kg/day) for ages 6-12 months. For ages 1-10 years the requirements for growth were based on the assumption that new tissue contains 30 μg/g (0. 46 mmol zinc/g). For adolescent growth, a zinc content of 23 μg/g (0. 35 mmol/g) increase in body weight was assumed. Pubertal growth spurts increase physiologic zinc requirements substantially. Growth of adolescent males corresponds to an increase in body zinc requirement of about 0. 5 μg/day (7. 6 mmol/day).

Pregnancy

The total amount of zinc retained during pregnancy has been estimated to be 1. 5 mmol (100 mg). During the third trimester the physiologic requirement of zinc is approximately twice as high as that in women who are not pregnant.

Lactation

Zinc concentrations in human milk are high in early lactation, 2-3 mg/l (31-46 mmol/l) in the first month, and fall to 0. 9 mg/l (14 mmol/l) after 3 months. From data on maternal milk volume and zinc content, it was estimated that the daily output of zinc in milk during the first 3 months of lactation could amount to 1. 4 mg/day (21. 4 mmol/l), which would theoretically triple the physiologic zinc requirements in lactating women compared with non-lactating, non-pregnant women. In setting the estimated requirements for early lactation it was assumed that part of this requirement was covered by postnatal involution of the uterus and from skeletal resorption.

Elderly

Requirements for the elderly are estimated to be the same as those for other adults. A lower absorptive efficiency has been reported in the elderly, which could justify a higher dietary requirement. On the other hand, endogenous losses seem to be lower in the elderly. Because of the suggested role of zinc in infectious diseases, an optimal zinc status in the elderly could have a significant public health effect and is an area of zinc metabolism requiring further research.

Inter-individual Variations in Zinc Requirements and Recommended Nutrient Intakes

The average physiologic requirements with the factorial technique have considered a relatively small number of subjects and do not allow any estimate of inter-individual variations in obligatory losses of zinc at different intakes. Because zinc requirements are related to tissue turnover rate and growth, it is reasonable to assume that variations in physiologic zinc requirements are of the same magnitude as variations in protein requirements and that the same figure (12. 5 percent) for the inter-individual coefficient of variation (CV) could be adopted. However, the estimates of dietary zinc requirements involve an estimate of absorption. Consequently, variations in absorptive efficiency, not relevant in relation to estimates of protein requirements, may have to be taken into account in the estimates of the total inter-individual variation in zinc requirements.

Systematic studies of the inter-individual variations in zinc absorption under different conditions are few. In small groups of healthy well-nourished subjects, the reported variations in zinc absorption from a defined meal or diet are of the order of 20-40 percent and seem to be independent of age, sex, or diet characteristics.

How much these variations, besides being attributable to methodologic imprecision, reflect variations in physiologic requirement, effects of preceding zinc intake, etc. is not known. From the available data from zinc absorption studies it is tentatively suggested that the variation in dietary zinc requirements, which covers variation in requirement for absorbed zinc (i. e., variations in metabolism and turnover rate of zinc) and variation in absorptive efficiency, corresponds to a CV of 25 percent. The recommended nutrient intakes derived from the estimates of average individual dietary requirements with the addition of 50 percent (2 standard deviations).

UPPER LIMITS OF ZINC INTAKE

Only a few occurrences of acute zinc poisoning have been reported. The toxicity signs are nausea, vomiting, diarrhoea, fever, and lethargy and have been observed after ingestion of 4-8 g (60-120 mmol) zinc. Long-term zinc intakes higher than the requirements could, however, interact with the metabolism of other trace elements. Copper seems to be especially sensitive to high zinc doses. A zinc intake of 50 mg/day (760 mmol) affects copper status indexes, such as CuZn-superoxide dismutase in erythrocytes. Low copper and ceruloplasmin levels and anaemia have been observed after higher zinc intakes 450-660 mg/day (6. 9-10 mmol/day). Changes in serum lipid pattern and in immune response have also been observed in zinc supplementation studies. Because copper also has a central role in immune defence, these observations call for caution before large-scale zinc supplementation programmes are undertaken. Any positive effects of zinc supplementation on growth or infectious diseases could be disguised or counterbalanced by negative effects on copper-related functions.

The upper level of zinc intake for an adult man is set at 45 mg/day (690 mmol/day) and extrapolated to other groups in relation to basal metabolic rate. For children this extrapolation means an upper limit of intake of 23-28 mg/day (350-430 mmol/day), which is close to what has been used in some of the zinc supplementation studies. Except for excessive intakes of some types of seafood, such intakes are unlikely to be attained with most diets. Adventitious zinc in water from contaminated wells and from galvanised cooking utensils could also lead to high zinc intakes.

The risk for inadequate zinc intakes in children has been evaluated by using the suggested estimates of zinc requirements and by using data available on food composition and dietary intake in different parts of the world. For this assessment it was assumed that the distribution of zinc requirements is Gaussian with a CV of 15 percent and that the correlation between intake and requirement is very low. Zinc absorption from diets in Malawi, Kenya, Mexico, and Guatemala was estimated to be 15 percent based on the high phytate-zinc molar ratio in these diets, whereas an absorption of 30 percent was assumed for diets in Ghana, Guatemala, Egypt, and Papua New Guinea.

Fermented maise and cassava products (kenkey, banku, and gari) in Ghana, yeast leavened wheat-based bread in Egypt, and the use of sago with a low phytate content as the staple in the New Guinean diets were assumed to result in a lower phytate-zinc molar ratio and a better availability. With this approach 68-94 percent of the children were estimated to be at risk for zinc deficiency in these populations, with the exception of Egypt where the estimate was 36 percent. The average daily zinc intakes of these children were 3. 7-6. 9 mg (56-105 mmol). Most of the zinc supplementation studies have not provided dietary intake data, which could be used to identify the zinc intake critical for growth effects. In a recent study in Chile, positive effects on height gain in boys after 14 months of zinc supplementation was noted. The intake in the placebo group at the start was 6. 3 ± 1. 3 mg/day (96 ± 20 mmol/day) (n=49). Because only 15 percent of the zinc intake of the Chilean children was derived from flesh foods, availability was assumed to be relatively low.

No effect of zinc supplementation on human-milk zinc content or on maternal status of a group of lactating women and judged their intake sufficient to maintain adequate zinc status through 7 months or more of lactation. The mean zinc intake of the non-supplemented women was 13. 0 ± 3. 4 mg/ day (199 ± 52 mmol/day).

The presumed negative impact of a high-phytate diet on zinc status, were confirmed in recent experimental studies. Reductions in urinary and faecal losses maintained normal plasma zinc concentrations over 5 weeks in 11 men with intakes of 2. 45 mg zinc/day (37 mmol/day) or higher in a diet with a presumably high availability. In a similar repletion-depletion study with 15 men, an intake of 4 mg/day (61 mmol/ day) from a diet with a molar phytate-zinc ratio of 58 for 7 weeks resulted in a reduction of urinary zinc excretion from 0. 52 ± 0. 18 to 0. 28 ± 0. 15 mg/day (7. 9 ± 2. 8 mmol/day to 4. 3 ± 2. 3 mmol/day).

A significant reduction of plasma zinc concentrations and changes in cellular immune response were observed. Effects on immunity were also observed when a zinc-restricted diet with a high phytate content (molar ratio approximately 20) was consumed by five young male volunteers for 20-24 weeks. Sub-optimal zinc status has also been documented in pregnant women consuming diets with high phytate-zinc ratios (>17). Frequent reproductive cycling and high malaria prevalence seemed to contribute to the impairment of zinc status.

FOLATE AND FOLIC ACID

ROLE OF FOLATE AND FOLIC ACID IN HUMAN

Folates accept one-carbon units from donor molecules and passes them on via various biosynthetic reactions. In their reduced form cellular folates function conjugated to a polyglutamate chain. These folates are a mixture of unsubstituted polyglutamyl tetrahydrofolates and various substituted one-carbon forms of tetrahydrofolate.

The reduced forms of the vitamin, particularly the unsubstituted dihydro and tetrahydro forms, are unstable chemically. They are easily split between the C-9 and N-10 bond to yield a substituted pteridine and p-aminobenzoylglutamate, which have no biologic activity. Substituting a carbon group at N-5 or N-10 decreases the tendency of the molecule to split; however, the substituted forms are also susceptible to oxidative chemical rearrangements and, consequently, loss of activity. The folates found in food consist of a mixture of reduced folate polyglutamates.

Although natural folates rapidly lose activity in foods over periods of days or weeks, folic acid (e. g., in fortified foods) is almost completely stable for months or even years. The chemical lability of all naturally occurring folates results in a significant loss of biochemical activity during harvesting, storage, processing, and preparation.

Half or even three-quarters of initial folate activity may be lost during these processes. This is in contrast to the stability of the synthetic form of this vitamin, folic acid. In this form the pteridine (2-amino-4-hydroxypteridine) ring is not reduced, rendering it very resistant to chemical oxidation. However, folic acid is reduced in cells by the enzyme dihydrofolate reductase to the di- and tetrahydro forms. This takes place within the intestinal mucosal cells, and 5-methyltetrahydrofolate is released into the plasma.

Natural folates found in foods are all conjugated to a polyglutamyl chain containing different numbers of glutamic acids depending on the type of food. This polyglutamyl chain is removed in the brush border of the mucosal cells by the enzyme folate conjugase, and folate monoglutamate is subsequently absorbed.

The primary form of folate entering human circulation from the intestinal cells is 5-methyltetrahydrofolate monoglutamate. This process is, however, limited in capacity. If enough folic acid is given orally, unaltered folic acid appears in the circulation, is taken up by cells, and is reduced by dihydrofolate reductase to tetrahydrofolate

The bio-availability of natural folates is affected by the removal of the polyglutamate chain by the intestinal conjugase. This process is apparently not complete, thereby reducing the bio-availability of natural folates by as much as 25-50 percent. In contrast, synthetic folic acid appears to have a bio-availability of close to 100 percent.

The low bio-availability and - more importantly - the poor chemical stability of the natural folates has a profound influence on the development of nutrient recommendations. This is particularly true if some of the dietary intake is in the synthetic form, folic acid, which is much more stable and bio-available. Food fortification of breakfast cereals, flour, etc. can add significant amounts of folic acid to the diet.

Functional folates have one-carbon groups derived from several metabolic precursors (e. g., serine, N-formino-L-glutamate, folate, etc.). With 10-formyltetrahydrofolate the formyl group is incorporated sequentially into C-2 and C-8 of the purine ring during its biosynthesis. Likewise the conversion of deoxyuridylate into thymidylate (a precursor to DNA) is catalysed by thymidylate synthase, which requires 5,10-methylenetetrahydrofofate. Thus, folate in its reduced and polyglutamylated forms is essential for the DNA biosynthesis cycle.

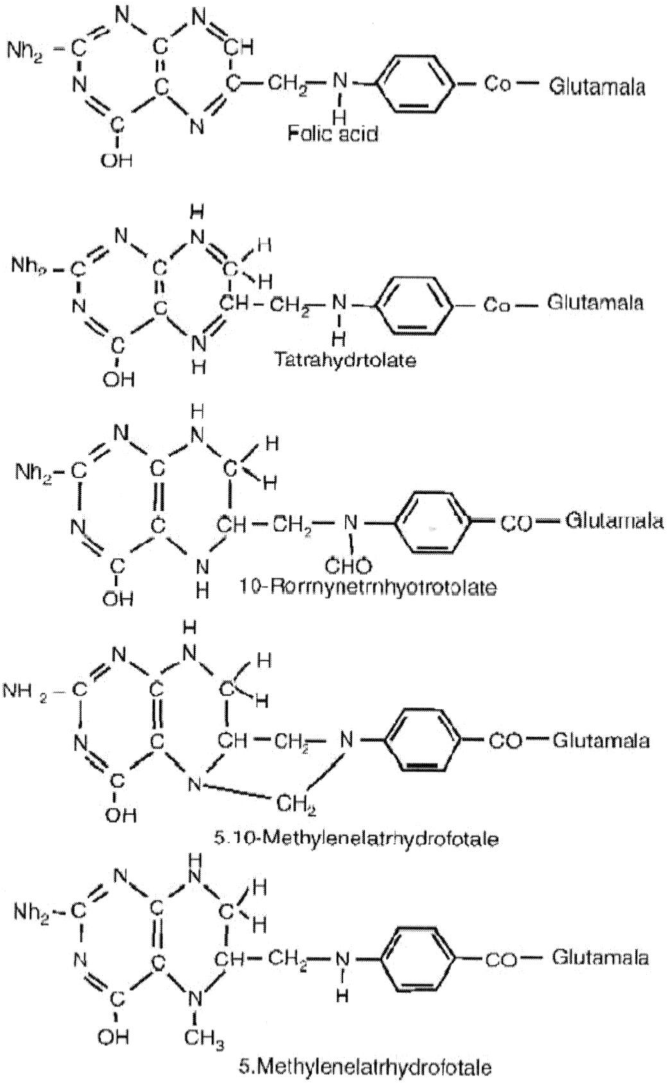

Fig. The chemical formula of folic acid (synthetic form) and the most important natural folates.

Alternatively 5,10-methylenetetrahydrofolate can be channelled to the methylation cycle. This cycle has two functions. It ensures that the cell always has an adequate supply of S-adenosylmethionine, an activated form of methionine, which acts as a methyl donor to a wide range of methyltransferases. These enzymes methylate a wide range of substrates including lipids, hormones, DNA, proteins, etc. One such important methylation is that of myelin basic protein, which acts as insulation for nerves cells. When the methylation cycle is interrupted as it is during vitamin B_{12} deficiency, one of the clinical consequences is the demyelination of nerve resulting in a neuropathy which leads to ataxia, paralysis, and, if untreated, ultimately death. Other important methyltransferase enzymes down-regulate DNA and suppress cell division.

In the liver the methylation cycle also serves to degrade methionine. Methionine is an essential amino acid in Humans and is present in the diet of people in developed countries at about 60 percent over that required for protein synthesis and other uses. The excess methionine is degraded via the methylation cycle to homo-cysteine, which can either be catabolised to sulfate and pyruvate (with the latter being used for energy) or remethylated to methionine. The need to maintain intracellular S-adenosylmethionine levels is related to the amount of methionine metabolised via homo-cysteine.

The DNA and methylation cycles both regenerate tetrahydrofolate. However, there is a considerable amount of catabolism of folate and a small loss of folate via excretion from the urine, skin, and bile. There is a need to replenish the body's folate content by uptake from the diet. If there is inadequate dietary folate, the activity of both the DNA and the methylation cycles will be reduced. A decrease in the former will reduce DNA biosynthesis and thereby reduce cell division. Although this will be seen in all dividing cells, the deficiency will be most obvious in cells that are rapidly dividing, for example, in a decrease in red cell production, producing anaemia. Other cells derived from bone marrow also decrease, leading to leucopenia and thrombocytopenia. Likewise there is a reduction in cell division in the lining of the gut. Taken together, this reduction in the DNA cycle results in an increased susceptibility to infection, a decrease in blood coagulation, and secondary malabsorption. In folate deficiency, the flux through the methylation cycle is decreased but the DNA cycle may be more sensitive. The most obvious expression of the decrease in the methylation cycle is an elevation in plasma homo-cysteine. This is due to a decreased availability of new methyl groups provided as 5-methyltetrahydrofolate, necessary for the remethylation of plasma homo-cysteine. Previously it was believed that a rise in plasma homo-cysteine was nothing more than a biochemical marker of possible folate deficiency. However, there is increasing evidence that elevations in plasma homo-cysteine are implicated in the aetiology of cardiovascular disease. This moderate elevation of plasma homo-cysteine occurs in subjects with a folate status previously considered adequate.

Interruption of the methylation cycle resulting from impaired folate status or deceased vitamin B_{12} or vitamin B_6 status may have serious long-term risks. Such interruption, as seen in vitamin B_{12} deficiency (e. g., pernicious anaemia), causes a very characteristic demyelination and neuropathy known as subacute combined degeneration of the spinal cord and peripheral nerves. If untreated, this leads to ataxia, paralysis, and ultimately death. Such neuropathy is not usually associated with folate deficiency but is seen if folate deficiency is very severe and prolonged. The explanation may lie in the well-established ability of nerve tissue to concentrate folate to a level of about five times that in the plasma. This may ensure that nerve tissue has an adequate level of folate when folate being provided to the rapidly dividing cells of the marrow has been severely compromised for a prolonged period. The resultant anaemia will thus inevitably present clinically earlier than the neuropathy.

DEFINITION OF POPULATIONS AT RISK FOR FOLATE DEFICIENCY

Nutritional deficiency of folate is common in people consuming a limited diet. This can be exacerbated by malabsorption conditions, including coeliac disease and tropical sprue. Pregnant women are at risk of folate deficiency because pregnancy significantly increases the folate requirement, especially during periods of rapid foetal growth (i. e., in the second and third trimester). During lactation losses of folate in milk also increase the folate requirement.

During pregnancy there is an increased risk of foetal neural tube defects (NTDs), with risk increasing 10-fold as folate status goes from adequate to poor. Between days 21 and 27 post-conception, the neural plate closes to form what will eventually be the spinal cord and cranium. Spina bifida, anencephaly, and other similar conditions are collectively called NTDs. They result from improper closure of the spinal cord and cranium, respectively, and are the most common congenital abnormalities.

DELINEATION OF DIETARY SOURCES

Although folate is found in a wide variety of foods, it is present in a relatively low density except in liver. Diets that contain adequate amounts of fresh green vegetables (i. e., in excess of three servings per day) will be good folate sources. Folate losses during harvesting, storage, distribution, and cooking can be considerable. Likewise, folate derived from animal products is subject to loss during cooking. Some staples, such as white rice and unfortified corn, are low in folate.

In view of the increased requirement for folate during pregnancy and lactation and by select population groups and in view of its low bio-availability, it may be necessary to consider fortification of foods or selected supplementation of women of child-bearing years.

DIFFERENCES IN BIO-AVAILABILITY OF FOLIC ACID AND FOOD FOLATE

The RNIs suggested for groups in Table used food folate as the source of dietary folate because most societies in developing countries consume folate from naturally occurring sources. As discussed in the introduction, natural folates are found in a conjugated form in food, which reduces its bio-availability by perhaps as much as 50 percent. In addition, natural folates are much less stable. If chemically pure folic acid (pteroyl monoglutamate) is used to provide part of the RNI, by way of fortification or supplementation, the total dietary folate, which contains conjugated forms (pteroyl polyglutamates), could be reduced by an appropriate amount.

On average the conjugated folate in natural foods is considered to be only half as available as synthetic folic acid. For example, the recommendation of usual mixed forms of folate in the diet is 400 μg/day, but 100 μg of this given as pure folic acid would be considered to be equivalent to 200 μg of dietary mixed folate. Hence, only an additional 200 μg of dietary folate would be needed.

NEURAL TUBE DEFECTS

It is now agreed that a supplement of 400 μg of folic acid taken near the time of conception will prevent most neural tube defects (NTDs). The recommendation to prevent recurrence in women with a previous NTD birth remains 4. 0 mg/day because of the high increase in risk in such cases and because that was the amount used in the most definitive trial. Because of the poorer bio-availability and stability of food folate, a diet based on food folate will not be optimum in prevention. One study determined that risk of NTD is 10-fold higher in people with poor folate status than in those with high normal folate status.

A further study suggests that an extra 200 mg/day or possibly 100 μg/day if taken habitually in fortified food would prevent most if not all of folate-preventable NTDs. Ideally, an extra 400 μg/day should be provided because this is the amount used in various intervention trials and can be achieved by supplementation. This amount could not be introduced by way of fortification, because exposure to high intakes of folic acid by people consuming a large intake of flour would run the risk of preventing the diagnosis of pernicious anaemia in the elderly. It is likely that depending on the staple chosen it would be possible to increase intake in most women by 100 μg/day without causing too high an exposure in other groups.

It is suggested that this amount, although not optimal, will prevent most NTDs.

CARDIOVASCULAR DISEASE

Plasma homo-cysteine concentration, if only moderately elevated, is an independent risk factor for cardiovascular disease and stroke. Increased risk was associated with values higher than 11 mmol/l, which is well within what is the normal range (5-15 mmol/l) of plasma homo-cysteine levels. In addition, even in populations that are apparently normal and consuming diets adequate in folate, there is a range of elevation of plasma homo-cysteine that could be lowered by an extra 100 or 200 μg/day of folic acid. Large-scale intervention trials regarding the significance of interrelationships among folate levels, plasma homo-cysteine levels, and cardiovascular disease have not been completed and therefore it would be premature to introduce public health measures in this area.

COLOURECTAL CANCER

Evidence suggests a link between colourectal cancer and dietary folate intake and folate status. One study reported that women who take multivitamin supplements containing folic acid for prolonged periods have a significantly reduced risk of colourectal cancer. However, the scientific evidence is not sufficiently clear for recommending increased folate intake in populations at risk for colourectal cancer.

UPPER LIMIT

There is no evidence that it is possible to consume sufficient natural folate to pose a risk of toxicity. However, this clearly does not apply to folic acid given in supplements or fortified foods. The main concern is the masking of the diagnosis of pernicious anaemia, because high levels of folic acid correct the anaemia, allowing the neuropathy to progress undiagnosed to where it may become irreversible even upon treatment with vitamin B_{12}. Consumption of large amounts of folic acid might also pose other less well-defined risks. Certainly, consumption of milligram amounts of folic acid would be undesirable.

Chapter 6

Dietary Antioxidants

NUTRIENTS WITH AN ANTIOXIDANT ROLE

The potential beneficial effects from antioxidants in protecting against disease have been used as an argument for recommending increasing intakes of several nutrients above those derived by conventional methods. If it is possible to quantify such claims, antioxidant properties should be considered in decisions concerning the daily requirements of these nutrients.

This section examines metabolic aspects of the most important dietary antioxidants - vitamins C and E, the carotenoids, and several minerals - and tries to define the populations which may be at risk of inadequacy to determine whether antioxidant properties per se should be and can be considered in setting a requirement. In addition, pro-oxidant metabolism and the importance of iron are also considered.

Members of the Food and Nutrition Board of the National Research Council in the United States, recently defined a dietary antioxidant as a substance in foods which significantly decreases the adverse effects of reactive oxygen species, reactive nitrogen species, or both on normal physiologic function in humans. It is recognised that this definition is somewhat narrow because maintenance of membrane stability is also a feature of antioxidant function and an important antioxidant function of both vitamin A and zinc. However, it was decided to restrict consideration of antioxidant function in this document to nutrients which were likely to interact more directly with reactive species.

The Need for Biologic Antioxidants

It is now well established that free radicals, especially superoxide ($O_2^{\cdot -}$), nitric oxide ($NO^{\cdot}$), and other reactive species such as H_2O_2, are continuously produced in vivo (5-7). Superoxide in particular is produced by leakage from the electron transport chains within the mitochondria and microsomal P450 systems or formed more deliberately, for example, by activated phagocytes as part of the primary immune defence in response to foreign substances or to combat infection by micro-organisms.

Nitric oxide is produced from L-arginine by nitric oxide synthases, and these enzymes are found in virtually every tissue of the mammalian body, albeit at widely different levels. Nitric oxide is a free radical but is believed to be essentially a beneficial metabolite and indeed it may react with lipid peroxides and function as an antioxidant. Nitric oxide also serves as a mediator whereby macrophages express cytotoxic activity against micro-organisms and neoplastic cells. If nitric oxide is at a sufficiently high concentration, it can react rapidly with superoxide in the absence of a catalyst to form peroxynitrite. Peroxynitrite is a potentially damaging nitrogen species which can react through several different mechanisms, including the formation of an intermediate with the reactivity of the hydroxyl radical.

To cope with potentially damaging reactive oxidant species (ROS), aerobic tissues contain endogenously produced antioxidant enzymes such as superoxide dismutase (SOD), glutathione peroxidase (GPx), and catalase and several exogenously acquired radical-scavenging substances such as vitamins E and C and the carotenoids. Under normal conditions the high concentrations of SOD maintains superoxide concentrations too low to allow the formation of peroxynitrite. It is also important to mention the antioxidant reduced glutathione (GSH). GSH is ubiquitous in aerobic tissues, and although it is not a nutrient, it is synthesised from sulfhydryl-containing amino acids and is highly important in intermediary antioxidant metabolism.

Integrated antioxidant defences protect tissues and are presumably in equilibrium with continuously generated ROS to maintain tissues metabolically intact most of the time. Disturbances to the system occur when production of ROS is rapidly increased, for example, by excessive exercise, high exposure to a xenobiotic compounds (such as an anaesthetic, pollutants, or unusual food), infection, or trauma. Superoxide production is increased by activation of NADPH oxidases in inflammatory cells or after the production of xanthine oxidase, which follows ischaemia. The degree of damage resulting from the temporary imbalance depends on the ability of the antioxidant systems to respond to the oxidant or pro-oxidant load.

Fruits and vegetables are good sources of many antioxidants, and it is reported that diets rich in these foods are associated with a lower risk of the chronic diseases of cancer and heart disease. Hence, it is believed that a healthful diet maintains the exogenous antioxidants at or near optimal levels thus reducing the risk of tissue damage. The most prominent representatives of dietary antioxidants are vitamin C, tocopherols, carotenoids, and flavonoids. Requirements for flavonoids are not being considered at this time and work on this subject is still very much in its infancy. In contrast, several intervention studies have been carried out to determine whether supplements of the other nutrients can provide additional benefits against such diseases.

The components in biologic tissues make an ideal mixture of substrates for oxidation. Polyunsaturated fatty acids (PUFAs), oxygen, and transition metals are present in abundance but are prevented from reaction by cellular organisation and structure. PUFAs are present in membranes but are always found with vitamin E. Transition metals, particularly iron, are bound to both transport and storage proteins; abundant binding sites on such proteins prevent overloading the protein molecule with metal ions.

Tissue structures, however, break down during inflammation and disease, and free iron and other transition metals have been detected. Potentially damaging metabolites can arise from interactions between transition metals and the ROS described above. In particular the highly reactive hydroxyl radical can be formed by the Fenton and Haber-Weiss reactions.

Pathologic conditions greatly increase the concentrations of both superoxide and nitric oxide, and the formation of peroxynitrite has been demonstrated in macrophages, neutrophils, and cultured endothelium. Peroxynitrite can react through several different mechanisms, including the formation of an intermediate with the reactivity of the hydroxyl radical.

$$Fe^2 + H_2O_2 = Fe^{3+} + OH + OH$$

$$O_2 + H_2O_2 = O_2 + OH + OH$$

$$NO + O_2 - = ONOO$$

During inflammation or other forms of stress and disease, new measures are adopted by the body to counter potential pro-oxidant damage. The body alters the transport and distribution of iron by blocking iron mobilisation and absorption and stimulating iron uptake from plasma by liver, spleen, and macrophages. Nitric oxide has been shown to play a role in the coordination of iron traffic by mimicking the consequences of iron starvation and leading to the cellular uptake of iron. The changes accompanying disease are generally termed the acute-phase response and are, generally, protective. Some of the changes in plasma acute-phase reactants which affect iron at the onset of disease or trauma.

Pro-oxidant Activity of Biologic Antioxidants

Most biologic antioxidants are antioxidants because when they accept an unpaired electron, the free radical intermediate formed has a relatively long half-life in the normal biologic environment. The long half-life means that these intermediates remain stable for long enough to interact in a controlled fashion with intermediates which prevent autoxidation, and the excess energy of the surplus electron is dissipated without damage to the tissues.

Thus it is believed that the tocopheroxyl radical formed by oxidation of a-tocopherol is sufficiently stable to enable its reduction by vitamin C or GSH to regenerate the quinol rather than oxidizing surrounding PUFAs. Likewise the oxidised forms of vitamin C, the ascorbyl free radical and dehydroascorbate, may be recycled back to ascorbate by GSH or the enzyme dehydroascorbate reductase. The ability to recycle these dietary antioxidants may be an indication of their physiologic essentiality to function as antioxidants.

Carotenoids are also biologic antioxidants but their antioxidant properties very much depend on oxygen tension and concentration. At low oxygen tension b-carotene acts as a chain-breaking antioxidant whereas at high oxygen tension it readily autoxidises and exhibits pro-oxidant behaviour. Palozza reviewed much of the evidence and suggests that b-carotene has antioxidant activity between 2 and 20 mmHg of oxygen tension, but at the oxygen tension in air or above (>150 mmHg) it is much less effective as an antioxidant and can show pro-oxidant activity as the oxygen tension increases. Palozza also suggests that autoxidation reactions of b-carotene may be controlled by the presence of other antioxidants (e. g., vitamins E and C) or other carotenoids.

There is some evidence that large supplements of fat-soluble nutrients such as b-carotene and other carotenoids may compete with each other during absorption and lower plasma concentrations of other nutrients derived from the diet. However, a lack of other antioxidants is unlikely to explain the increased incidence of lung cancer in the a-tocopherol b-carotene intervention study, because there was no difference in cancer incidence between the group which received both b-carotene and a-tocopherol and the groups which received one treatment only.

The free radical formed from a dietary antioxidant is potentially a pro-oxidant as is any other free radical. In biologic conditions which might deviate from the norm, there is always the potential for an antioxidant free radical to become a pro-oxidant if a suitable receptor molecule is present to accept the electron and promote the autoxidation. Mineral ions are particularly important pro-oxidants.

For example, vitamin C will interact with both copper and iron to generate cuprous or ferrous ions, respectively, both of which are potent pro-oxidants. Fortunately, mineral ions are tightly bound to proteins and are usually unable to react with tissue components unless there is a breakdown in tissue integrity. Such circumstances can occur in association with disease and excessive phagocyte activation, but even under these circumstances there is rapid metabolic accommodation in the form of the acute-phase response to minimise the potentially damaging effects of an increase in free mineral ions in extra-cellular fluids

Nutrients Associated With Endogenous Antioxidant Mechanisms

Both zinc and selenium are intimately involved in protecting the body against oxidant stress. Zinc combined with copper is found in the cytoplasmic form of SOD whereas zinc and magnesium occur in the mitochondrial enzyme. SOD occurs in all aerobic cells and is responsible for the dismutation of superoxide (reaction 4):

$$O_2^{\cdot -} + O_2^{\cdot -} + 2H^+ = H_2O_2 + O_2$$

Hydrogen peroxide produced as a product of dismutation reaction is removed by GPx of which selenium is an integral component (reaction 5). To function effectively, this enzyme also needs a supply of hydrogen, which it obtains from GSH. Cellular concentrations of GSH are maintained by the riboflavin-dependent enzyme glutathione reductase.

$$H_2O_2 + 2GSH = GSSG + 2H_2O$$

Four forms of selenium-dependent GPx have been described which have different activities in different parts of the cell. In addition, a selenium-dependent thyrodoxin reductase was recently characterised in human thyrocytes. Thyrodoxin reductase may be particularly important to the thyroid gland because it can cope with higher concentrations of peroxide and hydroperoxides generated in the course of thyroid hormone synthesis better than can GPx. It is suggested that in combination with iodine deficiency, the inability to remove high concentrations of hydrogen peroxide may cause atrophy in the thyroid gland, resulting in myxedematous cretinism.

SOD and GPx are widely distributed in aerobic tissues and, if no catalytic metal ions are available, endogenously produced superoxide and hydrogen peroxide at physiologic concentrations may have limited, if any, damaging effects. SOD and GPx are of fundamental importance to the life of the cell, and their activity is not readily reduced by deficiencies in dietary intake of these nutrients.

In contrast, enzyme activity can be stimulated by increased oxidant stress (e. g., ozone). Activities of zinc-dependent enzymes have been shown to be particularly resistant to the influence of dietary zinc, and although erythrocyte GPx activity correlates with selenium when the intake is below 60-80 mg/day, there is no evidence of impaired clinical function at low GPx activities found in humans. Nevertheless, one selenium intervention study reported remarkably lower risks of several cancers after 4. 5 years of selenium at 200 mg/day.

The effects were so strong on total cancer mortality that the study was stopped prematurely. However, the subjects were patients with a history of basal or squamous cell carcinomas and were not typical of the general population. In addition, a prospective analysis of serum selenium in cancer patients (1. 72 mmol/L) found very little difference from concentrations in matched controls (1. 63 mmol/L) although the difference was significant. Furthermore, areas with high selenium intakes have a lower cancer incidence than do those with low intakes, but the high selenium areas were the least industrialised.

Nutrients with Radical-quenching Properties

Vitamins C and E are the principal nutrients which possess radical-quenching properties. Both are powerful antioxidants, and the most important difference between these two compounds stems from their different solubility in biologic fluids. Vitamin C is water soluble and is therefore especially found in the aqueous fractions of the cell and in body fluids whereas vitamin E is highly lipophilic and is found in membranes and lipoproteins.

Vitamin E

Vitamin E falls into the class of conventional antioxidants which are generally phenols or aromatic amines. In the case of the four tocopherols that together constitute vitamin E, the initial step involves a very rapid transfer of phenolic hydrogen to the recipient free radical with the formation of a phenoxyl radical from vitamin E. The phenoxyl radical is resonance stabilised and is relatively unreactive towards lipid or oxygen. It does not therefore continue the chain.

However, the phenoxyl radical is no longer an antioxidant and to maintain the antioxidant properties of membranes, it must be recycled or repaired - that is, reconverted to vitamin E - because the amount of vitamin E present in membranes can be several thousand-fold less than the amount of potentially oxidizable substrate. Water-soluble vitamin C is the popular candidate for this role, but thiols and particularly GSH can also function in vitro.

There are eight possible isomers of vitamin E, but a-tocopherol (5,7,8 trimethyl tocol) is the most biologically important antioxidant in vivo. In plasma samples, more than 90 percent is present as a-tocopherol but there may be approximately 10 percent of g-tocopherol. In foods such as margarine and soy products the g form may be predominant and palm oil products are rich in the tocotrienols.

Vitamin E is found throughout the body in both cell and sub-cellular membranes. It is believed to be orientated with the quinol ring structure on the outer surface (i. e., in contact with the aqueous phase) to enable it to be maintained in its active reduced form by circulating reductants such as vitamin C. Within biologic membranes, vitamin E is believed to intercalate with phospholipids and provide protection to PUFAs. PUFAs are particularly susceptible to free radical-mediated oxidation because of their methylene-interrupted double-bond structure. The amount of PUFAs in the membrane far exceeds the amount of vitamin E, and the tocopherol-PUFAs ratios are highest in tissues where oxygen exposure is greatest and not necessarily where the PUFAs content is highest

Oxidation of PUFAs leads to disturbances in membrane structure and function and is damaging to cell function. Vitamin E is highly efficient at preventing the autoxidation of lipid and it appears as if the primary, and possibly only, role in biologic tissues is to provide this function. Autoxidation of lipid is initiated by a free radical abstracting hydrogen from PUFA to form a lipid radical which is followed by a rearrangement of the double-bond structure to form a conjugated diene.

In vitro the presence of minute amounts of peroxides and transition metals will stimulate the formation of the initial radical. Oxygen adds to the lipid radical to form a lipid peroxide (reaction 7) which then reacts with another lipid molecule to form a hydroperoxide and a new lipid radical. This process is shown in general terms below for the autoxidation of any organic molecule (RH), where the initial abstraction is caused by a hydroxyl radical ($OH^{\cdot}$).

$$RH + OH^{\cdot} = R^{\cdot} + H_2O$$

$$R^{\cdot} + O_2 = ROO$$

$$ROO^{\cdot} + RH = ROOH + R$$

Autoxidation or lipid peroxidation is represented by reactions. The process stops naturally when reaction between two radicals occurs more frequently than does reaction.

$$ROO^{\cdot} + ROO^{\cdot} = \text{non-radical products}$$

The presence of the chain-breaking antioxidant, vitamin E (ArOH), reacts in place of RH shown in reaction 8 and donates the hydrogen from the chromanol ring to form the hydroperoxide. The vitamin E radical ($ArO^{\cdot}$, tocopheroxyl radical) which is formed is fairly stable and therefore stops autoxidation. Hydroperoxides formed by lipid peroxidation can be released from membrane phospholipids by phospholipase A2 and then degraded by GPx in the cell cytoplasm.

$$ROO^{\cdot} + ArOH = ArO^{\cdot} + ROOH$$

Vitamin C

Many, if not all of the biologic properties of vitamin C are linked to its redox properties. For example, essential defects in scurvy such as the breakdown of connective tissue fibres and muscular weakness are both linked to hydroxylation reactions in which ascorbate maintains loosely bound iron in the ferrous form to prevent its oxidation to the ferric form, which makes the hydroxylase enzymes inactive. Ascorbate exhibits similar redox functions in catecholamine biosynthesis and in microsomal cytochrome P450 enzyme activity, although the latter may only be important in young animals.

In the eye, vitamin C concentrations may be 50 times higher than in the plasma and may protect against the oxidative damage of light. Vitamin C is also present in the gonads, where it may play a critical role in sperm maturation. Spermatogenesis involves many more cell divisions than does oogenesis, resulting in an increased risk of mutation. Levels of sperm oxidised nucleoside 8-OH-2'-deoxyguanosine (an indicator of oxidative damage to DNA) varied inversely with the intake of vitamin C (5-250 mg/day). No apparent effects on sperm quality were noted. Vitamin C was superior to all other biologic antioxidants in plasma in protecting lipids exposed ex vivo to a variety of sources of oxidative stress. The importance of vitamin C in stabilising various plasma components such as folate, homo-cysteine, proteins, other micronutrients, etc. has not been properly evaluated. When blood plasma is separated from erythrocytes, vitamin C is the first antioxidant to disappear.

Vitamin C is a powerful antioxidant because it can donate a hydrogen atom and form a relatively stable ascorbyl free radical. As a scavenger of ROS, ascorbate has been shown to be effective against the superoxide radical anion, hydrogen peroxide, the hydroxyl radical, and singlet oxygen. Vitamin C also scavenges reactive nitrogen oxide species to prevent nitrosation of target molecules. The ascorbyl free radical can be converted back to reduced ascorbate by accepting another hydrogen atom or it can undergo further oxidation to dehydroascorbate. Dehydroascorbate is unstable but is more fat soluble than ascorbate and is taken up 10-20 times more rapidly by erythrocytes, where it will be reduced back to ascorbate by GSH or NADPH from the hexose monophosphate shunt.

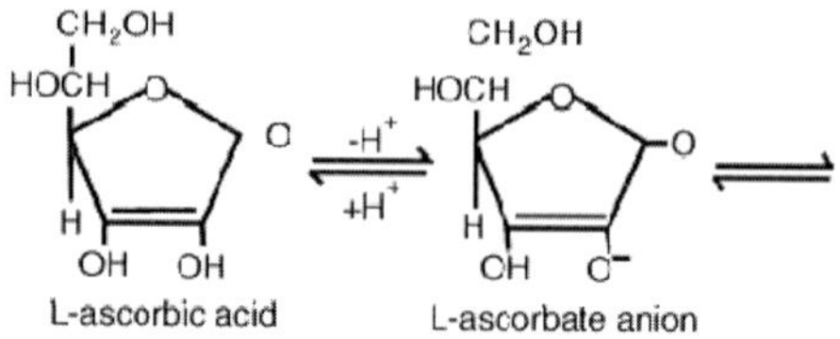

Fig. Vitamin C can donate a hydrogen atom and form a relatively stable ascorbyl free radical

Thus, mechanisms exist to recycle vitamin C similarly to those for vitamin E. The existence of a mechanism to maintain plasma ascorbate in the reduced state means that the level of vitamin C necessary for optimal antioxidant activity is not absolute because the turnover will change in response to oxidant pressure. Recycling of vitamin C will depend on the reducing environment which exists in metabolically active cells. In atrophic tissues or tissues exposed to inflammation, cell viability may fail and with it the ability to recycle vitamin C. In such an environment, the ability of newly released granulocytes or macrophages to scavenge vitamin C from the surrounding fluid may be invaluable for conservation of an essential nutrient as well as reducing the risk of ascorbate becoming a pro-oxidant through its ability to reduce iron.

xb-Carotene and Other Carotenoids

Many hundreds of carotenoids are found in nature but relatively few are found in human tissues, the five main ones being b-carotene, lutein, lycopene, b-cryptoxanthin, and a-carotene. b-Carotene is the main source of pro-vitamin A in the diet. There are approximately 50 carotenoids with pro-vitamin A activity, but b-carotene is the most important and is one of the most widely distributed carotenoids in plant species. Approximately 2-6 mg b-carotene is consumed by adults daily in developed countries and similar amounts of lutein and lycopene are probably also consumed. Smaller amounts may be consumed in the developing world. b-Cryptoxanthin is a pro-vitamin A carotenoid which is found mainly in fruits. Consumption is small but bio availability of carotenoids may be greater from fruit than vegetable, so its contribution to dietary intake and vitamin A status may be higher than the amount in the diet would predict.

b-Carotene has the general structure of this group of compounds and has two 6-membered carbon rings (b-ionone rings) separated by 18 carbon atoms in the form of a conjugated chain of double bonds. The latter is responsible for the antioxidant properties of the molecule, b-Carotene is unique in possessing two b-ionone rings in its molecule which are essential for vitamin A activity. The chemical properties of the carotenoids closely relate to the extended system of conjugated double bonds, which occupies the central part of carotenoid molecules, and various functional groups on the terminal ring structures. The ROS scavenged by carotenoids are singlet oxygen and peroxyl radicals. Carotenoids in general and lycopene specifically are very efficient at quenching singlet oxygen. In this process the carotene absorbs the excess energy from singlet oxygen and then releases it as heat. Singlet oxygen is generated during photosynthesis; therefore, carotenoids are important for protecting plant tissues, but there is limited evidence for this role in humans.

However, ß-carotene has been used in the treatment of erythropoietic protoporphyria, which is a light-sensitive condition which in some persons respond to treatment with amounts of b-carotene (in excess of 180 mg/day). It has been suggested that large amounts of dietary carotenes may provide some protection against solar radiation but results are equivocal. No benefit was reported when large amounts of b-carotene were used to treat persons with a high risk of non-melanomatous skin cancer. However, two carotenoids - lutein (3,3'-dihydroxy a-carotene) and seaxanthin (the 3,3'-dihydroxylated form of b-carotene) - are found specifically associated with the rods and cones in the eye and may protect the retinal pigment epithelium against the oxidative effects of blue light.

Burton and Ingold were the first to draw attention to the radical-trapping properties of b-carotene. Using in vitro studies, they showed that b-carotene was effective in reducing the rate of lipid peroxidation at the low oxygen concentrations found in tissues. Because all carotenoids have the same basic structure, they should all have similar properties. Indeed, several authors suggest that the hydroxy-carotenoids are better radical-trapping antioxidants than is b-carotene. It has also been suggested that because the carotenoid molecule is long enough to span the bilayer lipid membrane, the presence of oxy functional groups on the ring structures may facilitate similar reactivation of the carotenoid radical in a manner similar to that of the phenoxyl radical of vitamin E.

There is some evidence for an antioxidant role for b-carotene in immune cells. Bendich suggested that b-carotene protects phagocytes from auto-oxidative damage; enhances T and B lymphocyte proliferative responses; stimulates effector T cell function; and enhances macrophage, cytotoxic T cell, and natural killer cell tumoricidal capacity. Some data are in conflict with evidence of protective effects on the immune system and other data have found no effect. An explanation for the discrepancy may reside in the type of subjects chosen. Defences may be boosted in those at risk but it may not be possible to demonstrate any benefit in healthy subjects.

A Requirement for Antioxidant Nutrients

Free radicals are a product of tissue metabolism, and the potential damage which they can cause is minimised by the antioxidant capacity and repair mechanisms within the cell. Thus in a metabolically active tissue cell in a healthy subject with an adequate dietary intake, damage to tissue will be minimal and most of the damage occurring will be repaired. An important dietary source of antioxidant nutrients is the intake of fruit and vegetables, and it is now well established that persons consuming generous amounts of these foods have a lower risk of chronic disease than do those whose intake is small. These observations suggest that the antioxidant nutrient requirements of the general population can be met by a generous consumption of fruit and vegetables and the slogan "5 portions a day". Occasionally, damage may occur which is not repaired and the risk of this happening may increase in the presence of infection or physical trauma. Such effects may exacerbate an established infection or may initiate irreversible changes leading to a state of chronic disease (e. g., a neoplasm or atherosclerotic lesions). Can such effects also be minimised by a generous intake of dietary antioxidants in the form of fruit and vegetables or are supplements needed?

It is generally recognised that certain groups of people have an increased risk of free radical - initiated damage. Premature infants, for example, are at increased risk of oxidative damage because they are born with immature antioxidant status and this may be inadequate for coping with high levels of oxygen and light radiation. People who smoke are exposed to free radicals inhaled in the tobacco smoke and have an increased risk of many diseases. People abusing alcohol need to develop increased metabolic capacity to handle the extra alcohol load. Similar risks may be faced by people working in environments where there are elevated levels of volatile solvents (e. g., petrol and cleaning fluids, in distilleries, chemical plants, etc.). Car drivers and other people working in dense traffic may be exposed to elevated levels of exhaust fumes. Human metabolism can adapt to a wide range of xenobiotic substances, but metabolic activity may be raised with the consequent production of more ROS which are potentially toxic to cell metabolism.

Of the above groups, smokers are the most widely accessible people and this has made them a target for several large antioxidant-nutrient intervention studies. In addition, smokers often display low plasma concentrations of carotenoids and vitamin C. However, no obvious benefits to the health of smokers have emerged from these studies and, in fact, b-carotene supplements were associated with an increased risk of lung cancer in two separate studies and with more fatal cardiac events in one of them.

Other risk groups identified by their already having had some non-malignant form of cancer, such as non-melanomatous skin cancer or a colourectal adenoma, showed no effect on subsequent recurrences after several years of elevated intakes of antioxidant nutrients. The use of b-carotene or vitamin E alone or in combination with vitamin C showed no benefits. Thus, the results of these clinical trials do not support the use of supplementation with antioxidant micronutrients as a means of reducing cancer or even cardiovascular rates although in the general population, toxicity from such supplements is very unlikely.

Some intervention trials however have been more successful in demonstrating a health benefit. Stitch and colleagues gave large quantities of b-carotene and sometimes vitamin A to chewers of betel quids in Kerala, India, and to Canadian Inuits with pre-malignant lesions of the oral tract and showed reductions in leukoplakia and micronuclei from the buccal mucosa. Blot and colleagues reported a reduction (13 percent) in gastric cancer mortality in people living in Linxian Province, People's Republic of China, after a cocktail of b-carotene, vitamin E, and selenium. These studies are difficult to interpret because the subjects may have been marginally malnourished at the start and the supplements may have merely restored nutritional adequacy. However, correcting malnutrition is unlikely to be the explanation for the highly successful selenium supplementation study of US patients with a history of basal or squamous cell cancers of the skin. Interestingly, the intervention with 200 mg/day of selenium for an average of 4. 5 years had no effect on the recurrence of the skin neoplasms, confidence interval 0. 95-1. 28). However, analysis of secondary endpoints showed significant reductions in total cancer mortality and incidence and in the incidences of lung, colourectal, and prostate cancers.

The mean age of this group was 63 years and obviously they were not a normal adult population, but results of further studies are awaited with keen interest. Lastly, results of the Cambridge Heart Antioxidant Study should be mentioned because they provide some support for a beneficial effect of vitamin E in persons who have had a myocardial infarction. Recruits to the study were randomly assigned to receive vitamin E (800 or 400 mg/day) or placebo. Initial results of the trial suggested a significant reduction in non-fatal myocardial infarctions but a non-significant excess of cardiovascular deaths. The trial officially ended, but mortality has continued to be monitored and the authors now report significantly fewer deaths in those who received vitamin E for the full trial.

In conclusion, some studies have shown that health benefits can be obtained by some people with increased risk of disease from supplements of antioxidant nutrients. The amounts of supplements used have, however, been large and the effect possibly has been pharmacologic. Further work is needed to show whether more modest increases in nutrient intakes in healthy adult populations will delay or prevent the onset of chronic disease. The evidence available regarding health benefits to be achieved by increasing intakes of antioxidant nutrients does not assist in setting nutrient requirements.

ADULTERATION OF FOOD

Food Adulteration the act of adding or mixing something inferior, harmful, useless and unnecessary substance to food. In other words, any food item may be considered as adulterated if its nature and quality are not up to the standard. Unscrupulous traders normally adulterate food. In the process of adulteration, extraneous matters are directly added to food grains. Sands and crushed rock are added to increase weight. Mixing infested and damaged grains to good quality grains is a common practice. Sometimes grain polishing and husks are added to increase the weight. Nowadays, plastic beads that have the shape of food grains are often mixed with cereal grains. Coloured beads are added to the pulses. Sometimes water is sprayed over the grain stock to increase the weight.

Adulteration of fats and oils is easy and cannot be easily detected. *Ghee* (butter oil) is adulterated with hydrogenated oil and animal fats. Recently, because of the discovery of synthetic colours and flavours, any fat can be made to look like *ghee* and customers may easily be cheated. Til oil and coconut oil are often mixed with groundnut or cottonseed oil as the latter are cheaper. Mustard seeds are often mixed with argemone seeds and extracted together. Argemone oil contains an alkaloid-sanguinarine which is highly toxic and results in dropsy and paralysis. Adding allylisothiocyanate to soybean oil or palm oil gives the characteristic pungent smell of mustard oil. Mixing of palm oil with soybean oil is a common practice among dishonest traders for more profits.

The adulteration of milk is normally done with the addition of water and removal of fat. Sometimes extraneous substances like soybean and groundnut milk, wheat flour, etc are mixed. Selling diluted buffalo milk as cow milk is a common practice in rural areas. Addition of wheat flour, semolina, etc to milk powder is also common.

Tea leaves may be adulterated with the addition of used tea leaves, sawdust, and dried and ground leaves other than tea leaves. Spices like chillies and turmeric powder are adulterated with the addition of lead pigment to impart brightness in colour and good appearance. Metanil yellow, a carcinogenic agent, is used for colouring turmeric powder. chilli powder is normally adulterated by adding brick powder. Excessive use of wheat flour in place of milk protein (*chhana*) in the preparation of sweetmeat is an example of adulteration. Use of carboxy methyl cellulose (CMC) in lieu of liquid glucose or sugar syrup in the preparation of soft drinks is an example of extortion. In the name of various fruit juices, imitation products are prepared by using artificial and prohibited ingredients instead of using original fruit juice. Recently, a special drink named mineral water is being prepared and marketed with little or no assurance of quality.

Even this carefully-worded definition is not perfect. Some kinds of salt provisions have so much salt added that some of it has to be removed by soaking, to render the food edible, yet this does not constitute adulteration. Adulteration of food has long been practiced. It is mentioned in the case of bread by Pliny, who also says that difficulty was experienced in Rome in procuring pure wines. Athens had its public inspector of wines. England and France early passed laws to guard against the adulteration of bread, and as far back as the days of Edward the Confessor public punishment was provided for the brewers of bad ale.

The legal status of adulteration is largely a matter of statute, varying with each governmental body which attacks the subject. Food is declared adulterated if there is added to it a substance which depreciates or injuriously affects it; if cheaper or inferior substances are substituted wholly or in part for it; if any valuable or necessary constituent has been wholly or in part abstracted; if it is an imitation; if it is colored or otherwise treated, to improve its appearance; if it contains any added substance injurious to health.

These are examples of statutory provisions. Political considerations, such as the desire to protect the food-producers of a country, may affect legislation. Thus adulteration may be so defined as to include foreign products, which otherwise might be treated as unobjectionable. Food-preservatives have a very extensive use, which often constitutes adulteration. Salt is the classic preservative, but is also a condiment, and is seldom classed as an adulterant. Salicylic, benzoic, and boric acids, and their sodium salts, formaldehyde, ammonium fluoride, sulphurous acid and its salts are among the principal preservatives. Many of these appear to be innocuous, but there is danger that the continued use of food preserved by their agency may be injurious.

Extensive experiments on this subject have been performed by the United States Bureau of Chemistry and by the German Imperial Board of Health, among others. Some preservatives have been conclusively shown to be injurious when used for long periods, although their occasional use may be attended with no bad effect. Boric acid is pretty definitely condemned, after experiments on living subjects. Salicylic, sulphurous, and benzoic acids are indicated as injurious. The direct indictment against preservatives is not very strong. The principal point is that while the amount of preservative in a sample of food might be innocuous, the constant absorption of a preserving chemical by the system may have bad effects.

Preservatives are often sold for household use, as for the preparation of "cold process" preserves. If really made without heat, the tendency is, on the housekeeper's part, to use a proportion of the chemical larger than that employed by the manufacturer, thus increasing any bad effect attributable to them. Coloring matters are much used. Coal-tar colors are employed a great deal, and have received legal recognition in Europe. In the United States the tendency is rather to colour vegetable colors. Pickles and canned vegetables are sometimes colored green with copper salts; butter is made more yellow by anatta; turmeric is used in mustard and some cereal preparations. Apples are the basis for many jellies, which are colored so as to simulate finer ones.

This is an instance of the use of coloringmatter fraudulently, to imitate a more expensive article. But in confectionery dangerous colors, such as chrome yellow, Prussian blue, copper and arsenic-compounds are employed. Yellow and orange-colored candy is to be suspected. Fruit syrups, and wines, and tomato catsup are often artificially colored. Canned peas are especially to be suspected; often the fact that they are colored is stated on the label. Artificial flavouring-compounds are employed in the concoction of fruit syrups, especially those used for soda water. The latter are often altogether artificial. Among this class are: pear essence (amylic and ethylic acetates); banana essence (a mixture of amyl acetate and ethyl butyrate), and others. Milk is adulterated with water, and indirectly by removing the cream. It is also a favourite subject for preservatives. The latter are condemned partly because they render extreme cleanliness less necessary, for milk ordinarily exacts a high degree of purity in its surroundings. The addition of water may introduce disease germs. Cream is adulterated with gelatin, and formaldehyde is employed as a preservative for it. Butter is adulterated to an enormous extent with oleomargarine, a product of beef fat. It is a lawful product, but it is required by many enactments that its presence in butter be medicated on the package. Lard is another adulterant of butter.

Cheese is made from skim-milk some times, and cottonseed oil and other cheap fats are substituted for the cream. There are two principal sugar substitutes. One is glucose, with which sugar products are adulterated. It has less than two-thirds the sweetening power of sugar. The other is saccharine. This is the sweetest substance known; it is 230 times sweeter than sugar. It may be regarded as practically harmless. Sugar itself is generally pure. Meat is not much adulterated. It is generally only open to adulteration with preservatives, and cold storage causes these to be little used. It is sometimes dusted over with a preservative while in the piece, and sausages and similar products are often treated with preservatives and coloring matter. Boric acid and borax are typical preservatives, and sulphurous acid salts are used to restore a fresh appearance to stale meat. Starch is added to sausages. It is claimed that it prevents them from shrinking in cooking. Flour is adulterated by the addition of lower-grade meals, such as rye flour, corn meal, or potato starch; their use is not very common. Alum is employed to disguise the presence of damaged flour, and to prevent decomposition.

Alum is a still more frequent adulterant of bread; it is considered injurious to the animal system. Coffee is much adulterated, when sold ground. The root of chicory is a common adulterant, and even this has been supplanted by ocher and cheaper substances such as peas, beans, wheat, ground up after roasting. Attempts have been made to produce a counterfeit of the berry, an imitation being moulded out of some paste, but this has made no inroads. If coffee is bought unground, it will generally be pure, although the country of its origin may not be truthfully stated. Tea is generally pure, except that it may be of much lower grade than stated. Spent leaves are sometimes used, and the appearance is sometimes improved by "facing". This is the agitation with soapstone, Prussian blue, etc.

GOOD BACTERIA IN FOOD

Most people would think bacteria in or on food can only be harmful. True, food poisoning caused by bacteria and their products is a serious problem, and how to ensure food safety is treated in a different exhibit. However, certain bacteria are safe in food, and are required for the desired taste and texture. This exhibit describes some of the applications of bacteria and their products in the food industry.

Bacteria are commonly used in dairy products. Sour cream and Creme fresh (make your own! Creme fresh are both the products of cream after bacteria were allowed to grow in it. The difference in flavour, texture, and behaviour all result from the differences in bacteria required to produce the two products. Buttermilk is low in fat, cheese comes many variations. Yogurt is probably one of the oldest forms of fermented milk.

Is it not dangerous to eat food containing bacteria? Certainly not, as long as they are the right kind of bugs. It may even be benificial to eat diary products with living cultures, although the evidence is still controversial. True or not, fermented milk products are often suitable for people with a lactose intolerance though the product should not be heated after fermentation.

Fermented milk products are not completely fluid because casein, the major protein in milk, is insoluble in acid. Casein is also the protein that makes cheese solid. Fermented diary products taste different due to the bacteria used to make them. So what kind of bacteria are used for these foods? Most manufacturers will not release their magical strain of bacteria, producing just the right flavour, but here are their general characteristics:

- Table on diary bacteria and their products.
- What makes Limburger cheese so smelly? Isolate Brevibacterium linens.
- How Propionibacter shermanii makes the holes in Swiss cheese
- View images of the Yogurt bacteria L. bulgaricus.

Vegetables can be fermented too. Read about an ancient tradition. For instance, bacteria are the working lot to produce sauerkraut. The word 'sauerkraut' is German and means sour cabbage. And that is what it is: finely cut white cabbage leaves, left fermenting by lactic acid bacteria. Believe it or not: kraut only contains cabbage and salt. It doesn't taste salt because of the acid produced by the lactic acid bacteria, and the salt prevents fouling bacteria to grow. The bacteria used for sauerkraut is Leuconostoc mesenteroides. The same principle is used with pickling other foods, such as onions, olives, or meat. Almost always do lactic acid bacteria their little jobs to 'pickle to perfection'.

Have you ever thought that bacteria were used to produce chocolate and coffee? Although bacteria are not present in the final products containing chocolate, yeasts and bacteria degrade the pulp that covers fresh chocolate beans. Did you know that chocolate may not be bad for you in contrast to what your mother made you believe?.

Another example of bacteria helping to produce food products is vinegar. Whether good or bad depends on what the product should be. Wine makers will do their best to inhibit Acetobacter ferminting their grape juice: in order to produce wine, yeast should grow but not bacteria. However, if the desired product is vinegar, bacteria have their chance. Read more about wine and vinegar making in our special feature file.

Fermenting bacteria are used in the production of sausage. Again, lactic acid bacteria do their job. All you want to know about sausage. Don't worry, you can still enjoy your meat snacks since bacteria used in sausages are not at all related to E. coli or Salmonella, the enemy of every butcher. In fact, lactic acid bacteria are able to kill Listeria, a newly discovered ability that supports the safety of cured meats.

PATHOGENIC BACTERIA

Bacteria that cause disease are called pathogenic bacteria. Bacteria can cause diseases in humans, in other animals, and also in plants. Some bacteria can only make one particular host ill; others cause trouble in a number of hosts, depending on the host specificity of the bacteria. The diseases caused by bacteria are almost as diverse as the bugs themselves and include food poisoning, tooth ache, anthrax, even certain forms of cancer. It is impossible to sum up all bacterial diseases and it would be pretty boring. The Infectious Diseases fact sheets gives brief descriptions of diseases, including infectious diseases. Some diseases are named after the organisms that cause them, or is it the other way round?

Some pathogenic bacteria have received disproportionate attention in the press, e.g. the 'flesh-eating bacteria', which in real life are called Streptococci. Indeed they can cause spectacular, but fortunately uncommon symptoms. In the press, pathogenic bacteria are sometimes represented as (deadly) dangerous enemies that lurk in the dark, unseen, ready to attack you. Although that is exaggerated some bacteria can be life treatening, for example *Legionella pneumoniae,* the causative of Legionnaire's disease. These bacteria survive in moist places like air conditioners or hot-water pipes. Though potentially life treatening, bacterial infections do not kill all their victims. If they did, and if they spread fast, they would eradicate their hosts completely. Although a popular theme for thrillers, this is a 'mission impossible' in real life. Some bacteria kill a high percentage of people infected (they have a so-called a high mortality rate), but their relatively inefficient rate of spreading makes up for that.

When an infectious disease spreads around an area, and the cases of new infections reach a certain number, we call it an epidemic. If, on the other hand, a certain disease is always present in low numbers of cases in a given area, that disease is said to be endemic in that area. Some epidemics become wide-spread and quickly reach distant parts of the world: in modern times people travel fast and frequent, and our bacteria travel with us. A classical example of such a pandemic (in this case caused by a virus) was a new type of influenza ("Spanish 'flu") that reached continent after continent early this century, killing thousands of people on it's way. Although only a small proportion of the infected people died (the virus had a low mortality) so many people got infected that even the small proportion of deaths amounted to large numbers. An epidemic or pandemic can only occur if the population is not immune to that disease.

Read our special feature file on the history of infectious diseases about the times that pathogens could spread unlimited. Our exhibit on our immune system explains how immunity can prevent disease. So why do epidemics occur? Either because they are caused by diseases that did not exist before, like AIDS, caused by the HIV virus, or because new variants of bacteria (or viruses) arrive in an area where they were not endemic before.

This is why epidemics of the common 'flu' occur frequently: the virus causing influenza is able to change itself sufficiently to bypass our immunity built up from prior infections. Although not caused by bacteria, this site on influenza is very interesting if you want to understand epidemics of infectious diseases. Whenever a new virus type arises, there is the potential for a new epidemic. Fortunately, most of these new types are not as vicious as the Spanish 'flu. The experts are currently keeping an eye on bird flu, a strain of influenza that has not learned to jump from person-to-person. If that would happen, a new pandemic could be the result.

Bacteria have invented many different strategies to make us ill. These strategies, called bacterial pathogenicity, are the subject of an important division of medical microbiology. Understanding how certain bacteria make us ill can result in better treatment, vaccination, or prevention of that infectious disease. An example of such research is how treatment of ear infections can be improved. In another exhibit some of the common mechanisms of bacterial pathogenicity are explained. In order to keep this information balanced, now that you know what pathogenic bacteria are, why not also check out how commensal bacteria are good for you.

Index

A

Adolescents 38, 43, 47, 52, 53, 54, 57, 58, 65, 77, 87, 103, 109, 138, 141, 154, 157, 158, 160, 169, 170, 175, 176, 177, 181, 188, 197, 198, 203, 209, 217, 221, 223, 246, 247
Ageing 108, 159, 161, 190
Anatomy of The Digestive System 7
Atrophic Gastritis 82, 83

B

Biochemical Indicators 37, 40, 45, 49, 53, 56, 72, 73, 229

C

Calcium Absorption 146, 147, 148, 149, 150, 154, 155, 156, 158, 159, 160, 164, 167, 168, 170, 171
Calcium Intake 147, 148, 149, 150, 151, 153, 154, 155, 156, 160, 161, 162, 163, 165, 171
Carbohydrates 3, 8, 12, 37, 241
Cardiovascular Disease 31, 34, 116, 232, 256, 259
Colourectal Cancer 259
Criteria 130, 133, 201, 203, 217, 234, 235, 237

D

Deficiency 21, 22, 28, 29, 30, 31, 35, 36, 37, 38, 39, 40, 41, 43, 44, 47, 48, 50, 52, 55, 56, 62, 63, 66, 67, 71, 72, 73, 74, 75, 81, 82, 83, 84, 85, 87, 88, 89, 90, 91, 92, 93, 98, 102, 103, 104, 110, 114, 115, 117, 118, 119, 122, 124, 126, 127, 128, 129, 132, 133, 134, 135, 136, 137, 138, 140, 143, 146, 152, 155, 162, 164, 170, 172, 173, 174, 175, 177, 178, 179, 180, 182, 183, 190, 191, 194, 195, 196, 197, 198, 199, 200, 201, 202, 203, 204, 205, 206, 210, 211, 214, 215, 218, 219, 223, 226, 227, 228, 229, 230, 231, 233, 241, 242, 243, 245, 246, 251, 255, 256, 257, 267
Dietary Antioxidants 261, 263, 265, 274
Dietary Diversification 22, 23
Dietary Intake and Patterns 70

Dietary Sources 58, 64, 69, 79, 80, 95, 99, 117, 129, 174, 215, 219, 223, 243, 257
Digestive System 4, 5, 6, 7, 8

E

Elderly 22, 25, 30, 31, 42, 43, 46, 48, 51, 58, 77, 83, 87, 93, 103, 104, 106, 107, 109, 137, 143, 155, 162, 163, 190, 224, 232, 248, 249, 259
Ethnicity 107, 163
Excess Iodine Intake 182
Excretion 37, 38, 40, 41, 45, 49, 50, 53, 54, 56, 90, 98, 125, 127, 133, 136, 147, 149, 151, 154, 161, 164, 166, 170, 171, 172, 176, 177, 178, 180, 183, 215, 242, 243, 247, 252, 255

F

Factors Affecting Requirements 41, 46, 50, 53
Folate 24, 25, 26, 30, 84, 85, 90, 210, 252, 253, 254, 255, 256, 257, 258, 259, 271
Folic Acid and Food Folate 257
Folic Acid Fortification 30
Food Biotechnology 12
Food Processing 15, 16, 17, 18
Food Safety Testing 18
Food-Based Vitamins and Minerals 20
Fortification 26, 27, 28, 29, 30, 85, 109, 110, 182, 183, 188, 191, 210, 220, 227, 236, 244, 253, 257, 258

G

General Considerations 57

H

Health and Nutritional Benefits 13

I

Infants and Children 38, 43, 47, 51, 54, 57, 74, 77, 87, 135, 138, 157, 169, 187, 197, 203, 205, 221
Iodine Fortification 27, 28, 29, 31, 182, 183, 190, 191, 194, 195, 196, 197, 198, 199, 200, 201, 202, 203, 204, 205, 206, 210, 211
Iron and Zinc 25
Iron Fortification 27, 28, 210

K

Keshan Disease 228, 229, 230, 233, 236, 237

L

Lactation 39, 44, 47, 51, 52, 54, 55, 57, 64, 77, 86, 87, 88, 104, 105, 120, 138, 157, 160, 169, 181, 190, 206, 208, 222, 238, 239, 247, 248, 251, 257
Lipids 2, 12, 45, 59, 60, 112, 121, 241, 255, 271

M

Methodology 72, 88, 152
Microanatomy of the Digestive Tube 8
Microbial Life in the Digestive Tract 10
Morbidity and Mortality 63, 65, 74, 121

N

Neural Tube Defects 30, 257, 258
Niacin 42, 44, 45, 46, 47, 58

P

Pancreas 5, 8, 115
Pantothenate 52, 53, 54
Pathogenic Bacteria 283
Pernicious Anaemia 30, 82, 256, 258, 259
Pregnancy 39, 43, 44, 46, 47, 51, 52, 54, 55, 57, 64, 73, 76, 86, 87, 88, 104, 120, 138, 157, 159, 160, 169, 178, 181, 185, 188, 190, 196, 205, 206, 207, 208, 209, 210, 222, 238, 239, 242, 247, 248, 257
Pregnancy and Lactation 64, 86, 104, 120, 181, 190, 206, 238, 247, 257
Pro-Oxidant Activity 264, 265
Protein 8, 13, 15, 16, 20, 22, 45, 53, 58, 59, 60, 67, 68, 73, 89, 106, 112, 123, 124, 135, 145, 146, 148, 159, 163, 171, 172, 176, 185, 191, 192, 193, 194, 195, 196, 197, 200, 211, 219, 222, 224, 226, 229, 242, 243, 245, 246, 248, 249, 251, 255, 268, 274, 275, 276, 279, 284

R

Raw Materials 12, 15, 16, 102
Riboflavin 36, 39, 40, 41, 42, 43, 44, 46, 48, 58, 266

S

Selenium 224, 225, 226, 227, 228, 229, 230, 231, 232, 233, 234, 235, 236, 237, 238, 239, 240, 266, 267, 276
Supplementation 26, 27, 30, 31, 42, 95, 104, 117, 135, 137, 138, 139, 140, 159, 160, 168, 178, 190, 210, 222, 233, 235, 246, 250, 251, 257, 258, 275, 276

T

Thiamin 36, 37, 38, 58
Toxicity 37, 40, 48, 55, 78, 98, 109, 121, 142, 161, 250, 259, 275

U

Upper Limits 88, 160, 181, 223, 250

V

Vegetarians 28, 81, 88, 212

X

Xb-Carotene 272

Z

Zinc Fortification 29